NOISE IN
DIGITAL MAGNETIC
RECORDING

NOISE IN
DIGITAL MAGNETIC
RECORDING

Editors

T. C. Arnoldussen
L. L. Nunnelley

IBM Corporation
San Jose

World Scientific
Singapore • New Jersey • London • Hong Kong

Published by

World Scientific Publishing Co. Pte. Ltd.
P O Box 128, Farrer Road, Singapore 9128
USA office: Suite 1B, 1060 Main Street, River Edge, NJ 07661
UK office: 73 Lynton Mead, Totteridge, London N20 8DH

NOISE IN DIGITAL MAGNETIC RECORDING

ISBN 981-02-0865-0
      981-02-1025-6 (pbk)

Printed in Singapore by JBW Printers & Binders Pte. Ltd.

ACKNOWLEDGEMENTS

The editors wish to acknowledge the IBM Corporation for supporting this endeavor. In particular, they thank C. D. Mee for his active interest and very helpful advice. Among others who have aided the production of this book through managerial encouragement, topical discussions, reviews, or technical assistance are Gedeon Heinrich, Robert Rutledge, Dan Parker, Mason Williams, Gordon Hughes, and Mark Burleson. Special appreciation goes to the editors' wives, Christine Smith and Barbara Arnoldussen. Speaking on behalf of the contributing authors, the editors wish to thank their respective families and organizations for vital help and cooperation. Finally, acknowledging that the present understanding of noise in digital magnetic recording came about through the diligent and creative work of scores of researchers during the past decade and before, the editors hope that bringing this accumulated knowledge together here will provide a useful base for future progress.

CONTENTS

CHAPTER 1

INTRODUCTION

T. C. Arnoldussen and L. L. Nunnelley
IBM Corporation
San Jose, CA 95193 USA

Virtually every computer system, large and small, has a rigid disk system for magnetic data storage. In 1991 the total world production of rigid disks was approximately 100 million. The capability and capacity of these devices have improved enormously since their first development over thirty years ago. One chief measure of performance is surface data density, bits per area. Currently the surface density is increasing at about 30% per year. Increasing areal densities have permitted the use of smaller disk diameters while maintaining the same or higher data capacity per surface. Smaller diameters, in turn, permitted thinner disks without sacrificing rigidity. These factors combined to allow shrinking the entire disk drive, with volumetric storage density rising commensurately.

Prior to 1980, most magnetic storage for computers employed particulate recording media, iron-oxide particles in an organic binder. However, in the past decade thin film magnetic recording media have almost totally replaced particulate forms in rigid disk digital recording applications. Such disks employ substrates (usually aluminum alloy) with four or more thin films deposited by plating or sputtering. Only one of these is the magnetic recording layer, the others being used for mechanical purposes or for control of the magnetic layer properties. The process of building these structures is much more complex than that of coating disks with a particulate "paint."

Considerable thin film media research and development activity took place before 1980. During that time, efforts were directed at achieving controlled macroscopic magnetic properties, developing suitable mechanical underlayers and overcoats, and minimizing chemical corrosion. All the seminal ideas and technology existed to fabricate film disk media by the late 1970's, but the need did not yet exist. The 1980's brought personal computers, which, along with the ever increasing areal densities, motivated use of smaller substrates. 5-1/4 inch diameter disks were amenable to high throughput vacuum deposition, something rather unthinkable for the existing 14 inch substrates in use before. These changes caused the level of R&D to explode, and commercial manufacturing ensued. With film media being developed for real market applications, details of performance came under close scrutiny - noise taking a prominent place on the list of requirements. Media noise is one of the most important factors influencing system performance. For particulate media it wa generally well understood. Both the magnitude and the spectral shape is determined by the properties of the individual particles and the relatively weak interactions between them. In contrast, the noise from thin film media has proven to have a much richer assortment of mechanisms, magnitudes, and general behavior. The past decade has seen many measurements and theoretical treatments described in the literature, but not in a unified presentation.

This book surveys our present state of knowledge of noise in thin film media. Experimental information has come from two broad categories of activities. The first set of experimental activities involves measuring noise through a recording channel. The characteristics of the readback signal at the terminals of a head, flying on a disk, are examined. These measurements show that the noise in thin film media is associated spatially with magnetization reversals (written transitions) on the disk and not distributed uniformly along the track as is the case for particulate media. The second set of experimental activities has been the micromagnetic imaging of the written transitions on the medium to show the actual pattern of magnetization responsible for the noise. These images generally show that transitions have an irregular magnetization pattern, known as "zigzags."

Once the basic behavior of noise in thin film media was firmly established from the experimental work, theoretical efforts provided insight into the origins of media noise. Micromagnetic modeling techniques have been employed to explore a wider range of media properties and behavior than can be examined conveniently by experiment. Specifically the intergranular coupling can be turned on or off (independent of other effects) in a numerical model. This is seldom possible experimentally. These efforts have expanded our knowledge of thin film noise and have shown the relationship of noise characteristics to the detailed properties of the media. Once the noise mechanisms were clearly understood, the goal of manipulating the fabrication techniques to achieve very low noise thin film media became attainable. A natural outgrowth was to incorporate appropriate media noise behavior into models of system recording performance.

These topics and others are discussed at some length by separate authors in this volume. The chapters of this book are ordered so as to introduce the reader to the technology of film media materials and fabrication first, followed by a discussion of measured noise characteristics of film media. Chapter 2 (Kenneth Johnson) reviews film media materials and fabrication, primarily those aspects which affect media noise. In light of current understanding of film media noise, he discusses techniques for controlling intergranular magnetic coupling by alloy composition, process conditions, and film structure. Chapter 3 (Edward Murdock) discusses measurements which first revealed the unique nature of film media noise, as well as various later measurements used for characterization.

After these establishing chapters, the reader is led through the analysis of media noise and micromagnetic behavior. Chapter 4 (Thomas Arnoldussen) presents a global perspective of noise, based on statistical properties of physical fluctuations. A primary focus is on scale-dependent manifestations of noise. The conditions which produce additive noise and modulation noise are distinguished. Chapter 5 (Robert Ferrier) reviews state-of-the-art techniques for experimentally imaging micromagnetic features of films. Much of the early evidence that micromagnetic behavior controls recording and noise characteristics of film media came from Lorentz electron microscopy. The various Lorentz modes of microscopy are presented, as well as several new and promising imaging techniques, like SEMPA and MFM. The author points out strengths and limitations of different imaging tools. While media noise is the main focus of this book, methods for imaging micromagnetic structures of recording heads, as well as media, are included for completeness, because heads can also be a source of recording noise. Chapter 6 (Jian-Gang Zhu) makes use of dynamic micromagnetic computer modeling to explain the causes of various types of experimental images, examples of which appear in Chapter 5, and to show how micromagnetic interactions produce the observed recording noise behavior discussed in Chapter 3. The availability of supercomputers and massively parallel processing computers has made possible more detailed understanding of magnetic processes than ever before.

While thin film media noise is the newest form of noise in recording systems, other sources of noise and interference can also damage data reliability. To maintain a proper perspective, Chapter 7 (Edgar Williams) presents an overview of other recording system noise sources and degradation, pointing out their individual and collective effects. Methods for managing noise and interference, such as filtering and equalization, are described. Finally, Chapter 8 (Lewis Nunnelley) brings the reader's attention to some practical and important considerations of noise measurements. The theorist depends on the experimentalist for reliable characterization of real components, and the component and system engineers depend on both for proper design. Therefore the experimentalist must thoroughly understand his measuring instruments. Chapter 8 addresses crucial subtleties of noise measurements. Bandwidth and quantization noise, as well as techniques like zero-span frequency domain measurements and time interval analyses, are discussed.

Although terminology and symbols are relatively uniform across the recording field, there are occasional differences from author to author. Except where confusion would arise, the editors have allowed the contributors to this volume latitude to use their preferred terms and symbols of expression. For example, in one chapter "noise spectral density," NSD, is used while in another "noise power spectrum," NPS, appears. These are essentially the same quantity, but reflect different contextual emphases by different authors. Such variations not only allow the author flexibility, but expose the reader to synonymous terms which occur in the literature. In the index, however, such terms are grouped together under one term or concept, so the reader will be directed to appropriate, related areas of this book.

In a similar vein, magnetic units may vary from chapter to chapter. This truly reflects the usage prevalent in the field and it would be artificial and awkward to constrain authors to one system, like SI units. For theoretical treatments and for signal processing discussions SI units are convenient, or even preferred. However for discussion of magnetic properties of materials, it is far more common to use the cgs (Gaussian) system of units. Rather than making parenthetical translations, we will give the unit conversions and defining relationships here only. The pertinent quantities are magnetic field intensity (**H**), magnetic induction or magnetic flux density (**B**), magnetization (**M**), susceptibility (χ), and permeability (μ). μ_o is the permeability of free space. The defining relationships and units are summarized below.

SI	cgs (Gaussian)
$\mathbf{B} = \mu_o (\mathbf{H} + \mathbf{M})$	$\mathbf{B} = \mathbf{H} + 4\pi \mathbf{M}$
$\quad = \mu_o (1 + \chi)\mathbf{H}$	$\quad = (1 + 4\pi\chi)\mathbf{H}$
$\quad = \mu\mathbf{H}$	$\quad = \mu\mathbf{H}$

	SI	cgs (Gaussian)
B	Tesla, Webers/m^2	Gauss, Maxwells/cm^2
H	Amperes/m	Oersteds
M	Amperes/m	emu/cm^3
χ	dimensionless	dimensionless
μ,μ_o	Webers/Amp•m, Henries/m	dimensionless

The SI permeability of free space $\mu_o = 4\pi \ 10^{-7}$ Henries/m. **B**, **H**, and **M** may be converted from Gaussian to SI by the following relations.

$$B[\text{Tesla}] \quad = 10^{-4} \ B[\text{Gauss}]$$
$$H[\text{Amps/m}] \ = (10^3/4\pi) \ H[\text{Oersteds}]$$
$$M[\text{Amps/m}] \ = 10^3 \ M[\text{emu/cm}^3]$$

While the chapters here demonstrate that the understanding of thin film media noise has advanced considerably during the past decade, there remain areas of incomplete knowledge where more research is needed. When understanding is incomplete, there can be (and usually are) differences of opinion. In this volume, some authors offer mildly different interpretations of phenomena. Hopefully, these areas will indicate interesting outstanding problems.

This book, treating the specialized topic of noise, presumes the reader has some familiarity with magnetic recording. For more background in other aspects of magnetic recording technology, the reader is directed to the general references at the end of this introduction. None examine the field of media noise as extensively as this volume, but they represent excellent resources for this many faceted, multidisciplinary field.

References

Brown, W.F., *Micromagnetics*, Krieger, New York (1978).

Camras, M., *Magnetic Recording Handbook*, Van Nostrand Reinhold Company, New York (1988).

Hoagland, A.S. and J.E. Monson, *Digital Magnetic Recording*, Wiley-Interscience, New York (1991).

Jorgensen, F., *The Complete Handbook of Magnetic Recording*, TAB BOOKS, Inc., Blue Ridge Summit PA (1988).

Mee, C.D., *The Physics of Magnetic Recording*, North-Holland Publishing Company, Amsterdam (1964).

Mee, C.D., and E.D. Daniel, *Magnetic Recording Handbook: Technology & Applications*, McGraw-Hill   Book Company, New York (1990).

White, R.M., *Introduction to Magnetic Recording*, IEEE Press, New York (1985).

CHAPTER 2

FABRICATION OF LOW NOISE THIN-FILM MEDIA

KENNETH E. JOHNSON
IBM Corporation
Rochester, MN 55901

1.Introduction

Thin-film technology has replaced particulate technology as the primary method to fabricate rigid disks. The introduction of thin-film technology brought new challenges to the recording industry. A particularly important parameter is the noise characteristic of the metal films. Recording playback signals are adversely affected by the inherent noise present in a recorded track of a thin-film disk not observed using particulate technology. Thin-film media are recognized to have superior macromagnetic properties to other film technologies. The micromagnetics of thin-film media are just now becoming understood such that one can design a low noise media. This chapter will review the methods that have emerged to reduce noise in thin-film media.

The essence of a rigid disk is a magnetically hard film on a rigid substrate, and has evolved from magnetic particle dispersions coated onto rotating substrates forming polymer-particle composite structures to metal magnetic thin films capable of storage densities in excess of 1 Gbit/in^2 (Yogi et al., 1990b). Thin-film disks can be fabricated with the axis of magnetization either parallel or perpendicular to the surface of the substrate. Demagnetization arguments suggest that perpendicularly recorded bits can be recorded at higher densities than horizontal bits recorded with opposing polarities. This concept is true in principle, but head writing concerns, among others, have limited the advantage of perpendicular recording. Today's typical rigid disks use thin films having longitudinal anisotropies recorded at 100 Mb/in^2 on 95mm diameter substrates. It is expected that by the end of the decade storage densities in excess of 5 Gbit/in^2 using longitudinally recorded metal films deposited on 25mm diameter rigid disks shall have been accomplished.

One of the keys to obtaining storage densities of these magnitudes depends on the engineering of thin-film magnetic materials to form structures with high magnetization but very low noise. Thin-film disk noise was first recognized as having a different functionality vs. recording frequency, and to arise from different mechanisms than particulate technology in the early 1980s. Much research has been done in the last ten years to understand the interaction of thin-film processing variables with disk microstructures and micromagnetics to produce high signal to noise recording media. A fairly deep understanding of the interplay of thin-film disk processing, microstructure, and recording has been obtained. This understanding, particularly in regards to thin-film media noise reduction, was crucial to the 1 Gbit/in² demonstration in 1990 (Yogi et al., 1990b), and is just now being incorporated into current rigid disk products. This chapter will discuss the fabrication of rigid disk media, but with a focus on longitudinal thin-film media, and the methods and mechanisms of thin-film disk noise reduction.

2. Magnetic Property Overview

Rigid disk media magnetic properties can be conveniently subdivided into macroscopic and microscopic properties. Macroscopic properties such as coercivity (H_c), remanence-thickness product ($M_r\delta$), coercive squareness (S^*), and squareness (S) determine readback pulse shapes, amplitudes, and resolutions. These parameters are indicated in Fig. 1 which shows an M-H hysteresis curve taken with a vibrating sample magnetometer (VSM) of a thin-film disk suitable for readback with a magnetoresistive (MR) head. Microscopic properties such as grain or particle size, grain coupling, and grain crystallographic orientation determine the noise properties of the films. An optimization of macromagnetics for signal and micromagnetics for noise is necessary to make the optimum disk.

2.1 Macromagnetics

Theoretical work was presented in 1971 that allows one to calculate recording responses of media knowing H_c, M_r, δ, and S^* in conjunction with head fly height (d) (Williams and Comstock, 1971). From their analysis, the Williams-Comstock parameter a has been derived that relates macromagnetic properties of the disk, and field distribution of an inductive recording head to isolated pulse widths. The original expression has been simplified to the following, assuming S^* values close to one (Middleton, 1987):

$$a = \left[\frac{4M_r\delta(d + (\delta/2))}{H_c} \right]^{1/2} \tag{1}$$

Combining this factor with the formula for pulse width at 50% amplitude,

$$PW_{50} \cong \sqrt{g^2 + 4(d + a)(d + a + \delta)} \tag{2}$$

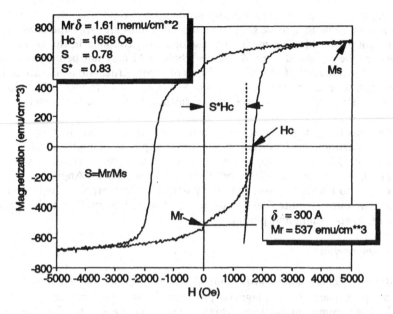

Figure 1. M-H hysteresis curve for a thin-film disk suitable for readback with an *MR* head. Film thickness is only 300 Å and the H_c value is in excess of 1600 Oe. This trend of decreasing thickness and increasing coercivity will continue for future thin-film disk constructs.

where g is the gap length of the head, one can approximate the readback voltage for any head-disk combination. The analysis has to be modified only slightly for the use of magnetoresistive (*MR*) heads. The general design criteria for rigid disk media is to minimize PW_{50} and a by increasing the coercivity H_c as high as possible without exceeding the writability of the head, keeping S^* high for narrow switching field distributions leading to narrow transitions, and keeping δ very low for high frequency output and to preclude MR head saturation. M_r has to be adjusted to give adequate signal output without sacrificing resolution.

The Williams-Comstock presentation and a subsequent analysis that took into account the large thickness of particulate coatings (Talke and Tseng, 1973), makes clear why a thin-film disk is the preferred recording medium. High H_c is easily attained in thin-film disks by using appropriate Co ternary and quaternary metal film alloys. Values in use today are between 1000 - 1500 Oe. The potential for H_c greater than 2400 Oe has been demonstrated. Particulate disks could conceivably achieve H_c values as high as 1500 Oe with the use of metal particles.

However, their use presents processing problems for rigid disk media. High $S*$ values are desirable to achieve small a values. $S*$ values are large and consistently greater than 0.8 in thin-film media because of strong coupling in the grain structure. As will be discussed later, $S*$ values above 0.90 may not be desirable since they are often indicative of high noise. Values of $S*$ for particulate media are usually less than 0.8. For this reason, the complete Williams and Comstock calculation, as opposed to using Eq. 1, is advisable for the calculation of the a parameter.

The greatest benefit of using thin-film media is rooted in how the $M_r\delta$ product is partitioned. Thin-film media are comprised of high moment metal materials with M_s values up to ten times that of a particulate coating. This high moment allows the use of films only several hundreds of Angstroms (Å) in thickness, up to ten times lower than the particulate counterpart. The fabrication of such thin films is quite feasible by using standard deposition techniques of sputtering, evaporation, and plating.

2.2 Micromagnetics

Both particulate and thin-film media magnetic properties are based on the aggregate properties of fine magnetic particles. Particulate rigid disk media are comprised of sub-micron particles, usually acicular in shape. Thin-film media are also comprised of "particles" or grains, but one tenth the size. Typical grain building blocks are 200-800 Å in diameter. In addition, the thin-film grains are much closer together, leading to stronger magnetostatic and exchange interactions. These stronger interactions have profound effects on both macromagnetics and micromagnetics.

Noise analysis in particulate media is in large part based on a particle volume packing density argument. Smaller particles allow a greater number density to be packed into a coating giving a statistical reduction to the noise. The particles do not interact by exchange and only slightly by magnetostatics. Noise is most prevalent in the dc erased state (net magnetization $= M_r$), where fluctuations in magnetization induced by particle orientation variation and particle flocculation causes variations in readback voltage. In a recorded track, the noise occurs uniformly along the recording direction.

In thin-film media, quite a different phenomenon is observed. Because of strong intergranular coupling, the grains tend to act cooperatively. In a dc erased state, regardless of varied preferred magnetic orientations of the grains, the magnetization direction over a large scale is constant. dc erase noise is low in thin-film media. On the other hand, when one investigates the actual transition region between oppositely magnetized regions in a recorded track, more chaotic behavior is observed. Zig-zag walls or sawtooth transitions are observed. The wall is irregular across the track width and this non-uniformity in the zig-zag leads to noise in the transition.

Noise in general has not been a particular problem for rigid disk recording products made in the 1980s, but new disk file products with recording densities in excess of 100 Mb/in² will require thin-film media with low noise. The understanding of the noise mechanisms and engineering of thin films to reduce strong coupling between grains while maintaining good macromagnetic properties has been an exciting development in the last ten years.

3. Disk Structure

3.1 Substrate

Figure 2 shows a cross sectional view of a rigid disk structure. The substrate of choice is aluminum because of its low density, rigidity, and low cost. The size of the substrate has continually shrunk since the first introduction. Over the past 35 years, there has been a stepwise evolution downward in diameters from 24″ to 65mm. Disk companies are today sampling substrate diameters of 48mm and lower for future products. Accompanying the shrinking diameters has been a reduction in substrate thickness. With the increase of areal densities on rigid disks has come the complementary increase of volumetric density, demanding thinner substrates for more efficient packing into drive designs. A substrate thickness standard of 1.91 mm lasted until the late 1980s. At this time, 1.27mm 130mm diameter and 0.80mm 95mm diameter disks were introduced. Future 65mm diameter products may have rigid disk thicknesses of 0.64mm, and future ceramic substrates may be as thin as 0.25mm.

A smooth surface is required for a rigid disk to give uniform readback signals from low flying heads. Aluminum by itself is quite soft and cannot be polished to a smooth enough finish. A technique of diamond turning has been used successfully for many years to produce mirror-like surfaces on Al substrates for particulate coatings. A stylus with a diamond tip cuts a spiral groove of several microns width continuously across the face of the disk giving a highly smooth and reflective surface. But this surface preparation technology is not adequate for high areal density thin-film disks. A modification of the Al substrate to give a hard surface for thin-magnetic film structural support, and to supply a surface capable of being polished to a high degree of smoothness for low flying recording heads was needed. Electroless nickel-phosphorus plating of Al is universally used today to supply this hard surface. Its hardness is measured at approximately 550 kg/mm² on the Brinell scale, almost equivalent to a cutting steel. The Ni-P surface is abrasively polished to a 20 Å root mean square (RMS) finish followed by a texturing process resulting in circumferential grooves measuring 50-100 Å RMS across the ridges. The grooves serve two purposes: the added

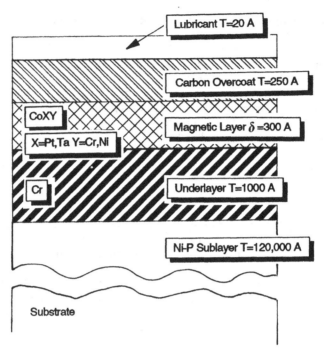

Figure 2. Cross section view of a thin-film rigid disk structure discussed in Sec. 3.

roughness minimizes head stiction to the disk surface, and the grooves induce a magnetic anisotropy circumferential on the disk, resulting in uniform magnetic readback signals during the course of a disk revolution.

Magnetic recording demands heads with low fly heights and narrow gaps for optimum areal density. Fly heights are limited by the finite roughness which is induced in the magnetic films by the Ni-P roughness. If tribological problems of stiction can be solved, disk substrate roughness can be reduced further. Ni-P surfaces are limited by abrasive polishing to about 10 Å RMS. The introduction of alternate substrate materials may be invoked to break this barrier. Ceramic and glass materials are already being evaluated as candidates for future rigid disk substrates. Ceramic materials offer many structural benefits, while glass gives the possibility of optically smooth 6 Å surfaces.

3.2 Magnetic Film Underlayer

Both particulate and thin-film media require underlayers, but for very different purposes. A chromate conversion coating has often been used as a sublayer for particulate coatings to prevent Al corrosion and to enhance magnetic film adhesion. Thin-film disks require an underlayer to help nucleate and grow microstructures giving appropriate magnetic properties. Cr is the most often used underlayer and is useful because of an epitaxial match of the Cr planes to preferred ones in the Co based magnetic alloys used in thin-film media. This epitaxy allows for alignment of the easy axes of magnetization in the plane or close to the plane of the disk for longitudinal media. Underlayers also present a microstructure that is replicated by the superimposed magnetic thin-film. As we shall see later, grain structures are critical in determining thin-film disk noise and the correct structure can be created in the magnetic film by supplying the right underlayer template.

3.3 Magnetic Film

Until recently, most rigid disk magnetic films consisted of magnetic particles dispersed in an organic binder. Today, rigid disk media are primarily made using thin-film technology and thus only thin-film technology will be considered in this chapter. There are several ways to deposit metal thin films. Electroplating and autocatalytic plating processes were initially investigated in the early 1950s (Bonn and Wendell, 1953; Brenner and Riddell, 1946), and in fact, the first thin-film disks for digital recording were introduced in the early 1960s using autocatalytic plating. Plating is still used today for the fabrication of thin-film disks, but it contributes only slightly to the entire volume of disks produced. It seems that plating techniques lack versatility, particularly in regards to attaining high coercivity and creating and controlling microstructure for low noise.

Vacuum deposition, or physical vapor deposition techniques, as opposed to solution deposition methods were first studied in the 1960s and 1970s (Bate, 1981). One can evaporate metallic thin films onto a substrate by heating a crucible with the desired material in a vacuum. Common heating methods are resistance heating and electron beam heating. Evaporation is an effective way to deposit large quantities of material rapidly, but lacks versatility in process control. It is a low energy process where the atoms arrive at the substrate with only thermal energies. Angles of incidence are critical to produce the desired magnetics and can vary greatly in an evaporation apparatus. Current thin-film disks consist of ternary Co alloys; controlling composition through evaporation is very difficult because of the different vapor pressures of the constituent. Thin-film disks have been experimentally produced using evaporation (Arnoldussen et al., 1984), however, sputtering has emerged as the primary method for the production of thin-film disks.

Vacuum deposition of magnetic thin films through sputtering was studied as early as the 1960s, but most of the work relative to thin-film disks was done in the 1980s (Howard, 1986; Eltouhky, 1986). Sputtering is accomplished by applying a voltage between the target material and the substrate to be sputtered in a vacuum vessel containing a sputtering gas. Argon is universally used because of its low cost and large atomic mass leading to good sputtering yields. A plasma of electrons and Ar ions is spontaneously generated upon voltage application and the Ar gas glows purple from the electronic excitations. Ar ions are accelerated onto the target material and by momentum transfer, atoms are displaced from the target and transferred to the substrate. The energy of the sputtered atoms is substantially higher than evaporated atoms and can aid in film adhesion. Thin film composition replicates the target composition in the sputtering process, allowing ease of use of binary and ternary magnetic alloys necessary for high coercivity low noise thin-film disks.

Sputtering can be done in several modes. Perhaps the simplest configuration is dc diode sputtering where a dc potential is placed between the target and substrate. Magnetron sputtering has been introduced to increase sputtering rates. In one common configuration, a magnetic field in the form of a racetrack is placed on the target by placing magnets on the target backside. Electrons are trapped close to the sputtering target increasing argon ion production and subsequent sputtering rate. Indeed, magnetron sputtering is the most common method to sputter thin-film disks. RF sputtering is also possible where a high frequency rf voltage is capacitively coupled to the target. RF sputtering allows for the deposition of non-conducting oxides and other materials which are possible candidates for thin-film disk overcoats and recording layers.

All the thin-film deposition techniques are effective in producing a uniform thickness magnetic film on the order of hundreds of Angstroms - the primary benefit of using thin-film media. Thin-film disks have magnetic layers spanning a thickness range from 200 to 1000 Å depending on the recording application. The thickness tolerance can be held to better than 5%. Such small thickness dimensions are not obtainable using particulate technology. A requirement of thin-film media is the necessity of a large magnetization (M_r) to balance the low thickness value in the $M_r\delta$ product. Fortunately, the Co alloys in use for thin-film media have M_s values between 500-800 emu/cm^3, values between 5-to-10 times the values of particulate coatings. Thin film materials have the benefit of intrinsically stronger magnetization than the magnetic oxides used in particulate media and, in addition, there is no dilution of magnetization from polymeric binders or non-magnetic particles often added for wear enhancement.

Materials used for the magnetic layer portion of thin-film disks are Co based binary and ternary alloys. Essentially three alloys are used in the production of current thin-film disks: CoCrTa, CoPtCr, and CoPtNi. Co is ferromagnetic and the source of large magnetocrystalline anisotropy giving the potential for large coercivities. Co by itself however, provides much too low a coercivity for recording densities required today. By forming alloys with large diameter transition elements such as Pt, Ta, Ir, and Sm, H_c values can be increased to values in excess of 2400 Oe (Howard, 1986; Kitada and Shimizu, 1984; Aboaf et al., 1983; Velu and Lambeth, 1991). Typical values of H_c are about 1000 Oe, but this value will increase as recording densities grow requiring narrower pulses (Sec 2.1).

Cr is a crucial second or third element in sputtered magnetic films and it serves two purposes. First, the presence of Cr in the alloy reduces corrosion potential tremendously. Plated CoP thin-film disks are particularly susceptible to corrosion, but it is the lack of Cr in the film and not the plating process that is to be blamed. It is thought that the Cr oxidizes and passivates the surface from further degradation. Secondly, the presence of Cr in the magnetic alloy allows for precipitation of alternate phases at grain boundaries or within grains that can aid in noise reduction (Sec. 5.2.1).

S and S^* values vary in thin-film media depending on the amount of grain interaction in the film. This interaction is controllable by adjusting process and composition variables. The discussion of these interactions will form a large part of the discussion of this article. Generally speaking, both squareness parameters are greater than values measured for particulate media and both are typically between 0.80 and 0.95. High S and S^* lead to large readback signals and sharply recorded transitions. However, there are increased noise implications for certain high S and S^* materials (See. Ch. 6) and thus, some thin-film disk manufacturers are making new disk products with reduced values of squareness.

3.3 Overcoats and lubricants

All thin-film disks have a sputtered protective overcoat; amorphous carbon has been the overwhelming choice of the thin-film disk industry. There have been some attempts at using a crystalline oxide zirconia overcoat (Yamashita et al., 1988). Overcoats are on the order of the thickness of the magnetic layer, and a 250 Å thickness is typical. The overcoat serves to protect the magnetic layer and also acts as a support structure for fluorocarbon lubricants (perfluoropolyethers).

4.Thin-Film Disk Noise

Theoretical treatment of particulate and film media noise is treated in Chapters 4 and 6. The reader is referred to Chapters 3 and 8 for detailed discussions of noise measurements. The rest of this chapter will discuss the origins of thin-film media noise and methods that have been developed to reduce thin-film media noise.

4.1 Thin-film Disk Microstructure

To understand the noise properties of thin-film disks, one must first understand the underlying structure of the magnetic film itself. Many TEM (transmission electron microscopy) studies have shown that the fundamental building blocks of a magnetic thin film are grains on the order of 100 - 500 Å in diameter. These grains can be separated from each other by up to 40 Å. Early work on plated CoP disks recognized the fundamental structure of small segregated grains, and suggested that the "channels" between the grains were composed of phosphorus (Aspland et al., 1969). Figure 3 shows several pictures of thin-film disk microstructures, all slightly different but all exhibiting a common grain unit structure. Grain size and grain crystallography are strongly affected by the underlayer structure - magnetic thin films generally replicate the microstructure of the underlayer. That is why underlayer studies concerning thickness, element type, alloy type, and crystallographic orientation have been so important to the development of thin-film disks. For very thick magnetic films, grain structure evolves independent of the underlayer, but for the first 500-1000 Å , the magnetic film takes on morphological and often crystallographic properties of the underlayer.

Electron and X-ray diffraction studies of rigid disk cobalt alloy thin-film media have consistently shown that the grains crystallize in the hcp (hexagonal close packed) structure with random distribution of their easy axes in two or three dimensions. Some authors have reported the presence of fcc phases in experimental thin films, but this phase seems to be rare and not apparent in present day commercial media. Since thin-film media are usually binary or ternary alloys containing larger atoms than the fundamental Co unit, the lattices are strained and this stress is observed experimentally by line shifts in the X-ray diffraction spectra. Up to a certain concentration, added metal elements enter substitutionally into the lattice with lattice stretching being proportional to size and concentration of the substituted atom. This strain may be responsible for the higher coercivities observed with binary alloys containing Pt, Ta, and other large atom additives.

The orientation of the crystallites in the structure is also important. In the hcp system, the easy axis of magnetization is along the c-axis. The distribution of the c-axes in the thin film affects S and S^* values and may relate to the detailed description of a transition (Sec. 4.3.4.1). Perpendicular thin-film media are purposely grown with c-axes perpendicular to the substrate. In longitudinal media, a partial perpendicular growth is often seen if a non-matching underlayer is used since growth along the (002) direction is favored. By manipulating the preferred orientations of the underlayer, the magnetic layer can be grown with the c-axes planar or close to planar.

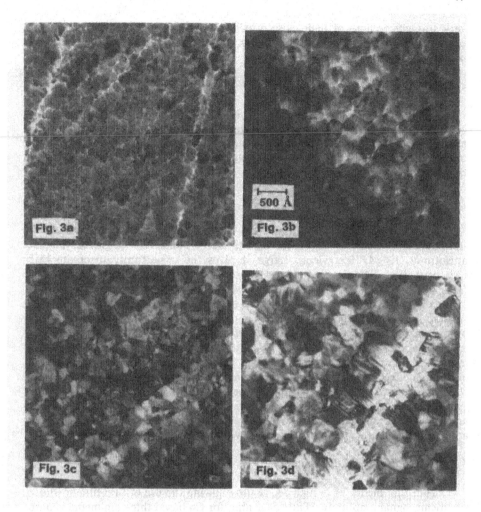

Figure 3. TEM micrographs of four different magnetic films. The fundamental structure is based on grains whose size can range from 100-500 Å. Intergranular spacing varies with different film deposition processes. Figures 3a and 3b show some evidence of grain separation. Figure 17b shows an extreme case of grain segregation through physical voiding.

The realization of a granular structure in thin-film disks has led theoreticians to base their efforts on the model of strongly interacting hcp single crystal grains, each having a uniaxial magnetocrystalline anisotropy. The grains interact very strongly because of their close proximity, making for difficult calculations, but ones that have yielded great insights into the mechanisms of coercivity and noise.

4.2 Thin-Film Disk Macromagnetics

The macromagnetic properties of thin films are well suited to advanced recording applications. Figure 1 refers to these properties on a hysteresis curve of a thin-film disk suitable for readback using a MR head. An understanding of the origin of these properties is necessary to understand the consequences of noise in thin-film media.

4.2.1 Coercivity (H_c)

The use of Co in thin-film rigid media is the natural choice of the ferromagnetic elements because of its large uniaxial magnetocrystalline anisotropy (K $= 4x10^6$ergs/cm³. Large K gives the opportunity to create high coercivity films. The detailed effects of process and microstructure variables on H_c is somewhat unclear. It has been shown by many workers that the addition of large atoms such as Pt, Ir, Sm, and Ta to the Co lattice increases the H_c value substantially. Grain size is also a factor in determining H_c. Thirty years ago, work was introduced presenting a curve of H_c vs grain size for different magnetic materials (Luborsky, 1961). Below a critical diameter, the grains become single domains and there is a coercivity maximum. Below this grain size, H_c drops because of thermal effects until H_c goes to zero at the superparamagnetic limit. This broad correlation between H_c and grain size seems to have been observed by many thin-film disk workers. A third contribution to the H_c mechanism is the interaction between adjacent grains. Well segregated structures where magnetostatic and exchange interactions are minimized will increase H_c.

4.2.2 Saturation Magnetization (M_s)

Thin-film media have high M_s values allowing the use of thin films without sacrificing signal amplitude. The upper limit for Co based thin-film media is pure Co with $M_s = 1722$ emu/cm³. Thin-film disk alloys have nonmagnetic constituents which reduce M_s. For the CoCr binary alloy, M_s decreases to zero at approximately 25% Cr. The situation is a bit more complicated for ternary alloys, but it is generally true that M_s is linearly related to Co atomic percent. The M_s-Co concentration functionality leads to a trade-off between magnetic moment in the film and the benefits of non-magnetic constituents that can improve coercivity and possibly aid in grain segregation (Sec.5.2). Some workers have attempted to overcome the loss in M_s from second element additions, by adding a third element, Ni, to form ternary alloys such as CoNiCr or CoPtNi. Higher H_c films in the range of 1000-1500 Oe can be obtained while maintaining M_s values of

800-1000 emu/cm³. Achieving H_c values beyond 1500 Oe is difficult, and requires non-ferromagnetic additives that decrease M_s.

4.2.3 Coercive Squareness (S*)

$S*$ has become a common way to evaluate the strength of coupling in thin films and in many cases the signal to noise characteristics. Very square hysteresis curves in any direction were measured on early CoP plated disks - a phenomenon quite different from particulate media. Particulate coating magnetic properties such as $S*$ and orientation ratios OR (ratio of M_r in the circumferential direction divided by M_r in the radial direction) are dictated by the uniaxial shape anisotropy of the ellipsoidal particles and the particle orientation distribution. For thin-film media, the exchange interactions are typically strong enough to dominate the uniaxial magnetocrystalline anisotropies of the individual grains. Once a highly interacting film is magnetized in any direction, the strong magnetostatic and exchange forces will keep the film magnetized in that same direction, regardless of grain size and easy axis distribution. OR values of 1 and $S*$ values in excess of 0.9 in any measured direction are typical for thin-film media when no macroscopic anisotropy is superimposed on the media by external process variables. If interactions are weakened by exercising process control, anisotropies can develop and $S*$ values will assume values in line with crystallographic and stress factors.

4.3 Thin-film disk micromagnetics

Simple mathematical models describe the average change in magnetization direction as an "S" type function such as an arctan, tanh, or error function. These functions ignore the details of the reversal and cannot describe the noise. Macromagnetics define the signal - micromagnetics define the noise. Optimizing noise through micromagnetic adjustments requires an experimental and theoretical understanding of what is happening at the transition.

4.3.1 Experimental observations

Early work (Dressler and Judy, 1974; Daval and Randet, 1970; Curland and Speliotis, 1970) using Lorentz microscopy showed the existence of a sawtooth-like magnetization pattern in various thin-film media. Figure 4a shows a schematic picture of such a zigzag transition. Certainly the amplitude of the zigzag will limit the linear density of the film. It has been shown that the length of the transition, as roughly measured by the peak to peak distance of the zigzag, correlates with the Williams-Comstock a parameter (Middleton and Miles, 1990; Lee and George, 1985). The fact that the transition is not constant across the track width as the transition evolves is not a factor for pulse height and width considerations. The problem with these zigzags comes from the fact that there are variations in the sawtooth. Figure 4b shows a more realistic situation; here the magnetization boundary does not follow a perfect sawtooth function, but rather undulates somewhat randomly across the track. This irregularity leads to noise in the transition. Since the noise is located within the transition region, the total noise is proportional to recording density (Baugh et al., 1983).

Zigzag transitions have now been observed by many workers using several methods. Techniques with resolution higher than the Bitter method are needed to reveal the transition detail. Lorentz microscopy was the first analysis that revealed the zigzag transition structures clearly (Daval and Randet, 1970; Dressler and Judy, 1970; Curland and Speliotis, 1970; Chen et al., 1978; Chen et al., 1981; Tong et al., 1984; Arnoldussen and Tong, 1986). Figure 5 shows a Lorentz micrograph of a zigzag transition recorded on a thin-film disk using a head flying 160-260 nm above the surface (Tong et al., 1984). Electron holography was used for the first time in 1983 (Yoshida et al., 1983) to reveal zigzag patterns on a Co evaporated thin film. The technique confirmed the existence of zigzag transitions, but is cumbersome for routine work. Recent developments have seen other analytical tools emerge that are useful for investigating the details of a thin-film disk magnetic transition. These are thoroughly discussed in Chapter 5.

Although many tools now exist that reveal the curious nature of the zigzag transition in thin-film disks, basic understanding of the mechanisms of transition structure in isotropic media and the general magnetic structures of thin films came from Lorentz microscopy work (Chen and Charlan, 1979; Chen, 1981). The Lorentz microscopic study of CoRe films showed the existence of magnetic clusters as evidenced by ripple structures separating one cluster from another. These clusters are groups of many magnetic grains all having the same magnetization direction because of strong coupling forces - the grains do not necessarily have the same magnetocrystalline easy axis direction. In the remanent state, the net magnetization of all the clusters is in one direction, but any one cluster can point in any given direction. When the film experiences a reversing field equal to the coercivity, the grains do not switch individually, but as clusters, leading to an avalanche change in magnetization. The high S^* value present in magnetic metal thin films is evidence for the presence of avalanche switching. The cluster becomes the basic unit of magnetization and Fig. 6 reproduces a schematic from Chen's paper describing the structure of thin film disk ripple as observed in CoRe. When remanent ripple structures are brought close together, a pattern develops as presented in Fig. 7. There is a change of magnetization direction across the transition involving a large angle rotation of the net magnetization direction of many groups of magnetic clusters forming vortex structures in the plane of the disk.

Thin-film media with strongly coupled grains exhibit transitions composed of aggregates of small grains acting as cluster units. Seemingly, cluster size could

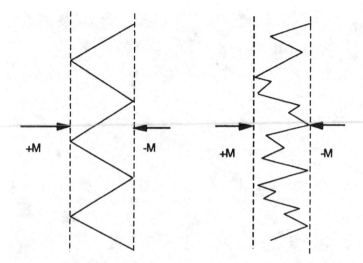

Figure 4. Schematic diagrams of zigzag transitions in thin-film media. The magnetization reverses direction through a transition whose width is the distance between the dotted lines. This width is governed by a parameter formulas. The structure within the dotted lines is governed by the micromagnetic properties. Figure 4a is an idealized picture of a zigzag transition where the sawtooth functionality is periodic. Figure 4b is a more realistic representation showing a more random structure.

become the limiting factor to transition lengths and overshadow simple $M_r\delta/H_c$ arguments, but data to date suggests that the simple macromagnetic approach to transition lengths still applies (Middleton and Miles, 1990). The existence of clusters, is the cause of noise in the transition regions. Magnetization cannot vary smoothly, but rather in discrete units.

4.3.2 Thin-film disk noise vs. recording density

In 1982, it was discovered that thin-film disks showed a noise vs. recorded frequency behavior very different from particulate disks reflecting the cluster model in the transition (Tanaka et al., 1982). Noise in CoNiP plated thin-film disks minimized in the dc saturation condition and increased with recording density. Gamma iron oxide particulate films had an opposite functionality where the noise is greater in the dc state. The now accepted interpretation is that noise in a contiguous plated film is found in the magnetic transitions, while noise in gamma iron oxide particulate media appears uniformly along the recording track.

Later recording experiments extended this work to other alloys and gave deeper insights into the noise vs recording density curves and the implied noise mechanisms (Baugh et al., 1983). Noise power in thin film alloys of CoNi, Co,

Figure 5. Lorentz micrograph of zigzag walls in recorded transitions on a thin-film disk. The sawtooth walls are not regular-the amplitude and pitch of the walls vary (Tong et al., 1984) © 1984-IEEE.

and CoRe were investigated to densities of 1500 fc/mm as shown in Fig. 8. As Tanaka et al. observed, the noise increased with recording density in a manner exactly opposite to particulate media and sputtered gamma iron oxide. The noise in the thin film media goes up linearly and then increases in slope in a supralinear region prior to reaching a maximum noise level. The noise behavior in the linear region is interpreted as simply the contributions of the extra transitions as the density is increased. The supralinear region indicates that the transitions become intrinsically noisier when they are close together. At very high densities when the transitions overlap, the noise power decreases. An important addition of this work is the suggestion that peak jitter can be predicted by signal to noise analysis, but only by using the maximum noise in the transition density curve.

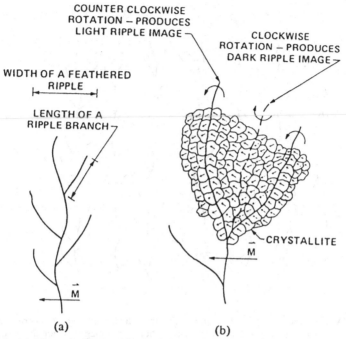

Figure 6. (a) Schematic drawing of a ripple image. (b) The schematic drawing shows the relation between cluster, crystallites and the ripple image. M represents the net magnetization direction. → The small arrow in the crystallite represents actual magnetization direction (Chen, 1981) © 1981-IEEE.

A proposed shifted transition model to explain the nature of thin-film disk noise vs transition density curves was proposed (Belk et al., 1985). Noise in general is associated with the random fluctuation of a variable. Belk et al. postulated that this variable is the location of the zigzag transition which is shifted from the expected position. Zigzag domains as noted above have been observed by several investigators using several techniques. A calculation of the RMS noise voltage based on shifted transitions reproduced the experimentally measured linear noise power vs transition density curves at low densities. By assuming a negative correlation (adjacent transition jitter tends to move in the same direction), the supralinear portion of the curve at higher densities is also simulated. The experimentally measured variance in transition shifts corresponds to the noise of the extended linear part of the noise vs. transition density curve at the density where the noise is maximum.

However, Lorentz microscopy work on actual recorded transitions (Arnoldussen and Tong, 1986) later showed that adjacent transitions were spaced

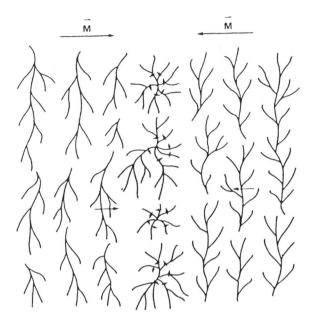

Figure 7. A schematic illustration of vortex structures at a transition and the corresponding magnetization rotation (The small arrow shows direction of net magnetization across ripple.) (Chen, 1981) © 1981-IEEE.

alternately closer and farther apart suggesting positive correlation. The Belk mechanism had to be modified. The current thinking is that positive correlation between adjacent bits results in domain bridging and ultimately island formation as the recording densities increase. The transitions are inherently more noisy at the high densities; the increased variance may come in the transition amplitude (Arnoldussen and Tong, 1986), or in the transition jitter induced from transition demagnetizing fields during the write process (Madrid and Wood, 1986; Barany and Bertram, 1987). This is further discussed in Chapter 3,4, and 6.

A popular method of measuring noise has emerged to complement the generation of noise vs. transition density curves. By first dc saturating a track on a disk in one direction, and then measuring the noise after the media is partially dc-erased in the other direction, a reverse dc erase noise curve is generated (Aoi et al., 1986; Bertram et al., 1986). The peak noise occurs when the head field plus demagnetizing fields in the medium are equal to the remanent coercivity H_r. The peak noise amplitude correlates with signal recorded media noise at high densi-

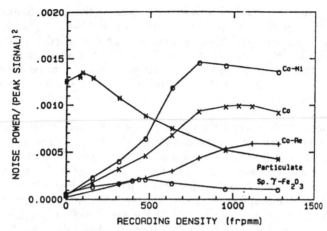

Figure 8. Total normalized medium noise power $\overline{N}^2$ vs. recording density for various material systems (Baugh et al., 1983) © 1983-IEEE.

ties. The technique is easy to implement and can rapidly give noise results on experimental media without the complications of performing spectral analysis.

A series of papers linking the reverse dc erase noise peak and fluctuations of certain disk parameters has been published (Bertram et al., 1986; Tarnopolsky et al., 1991; Tarnopolsky et al., 1989; Silva and Bertram, 1991). Sources of the fluctuations have been suggested to be large-scale variations in the coercivity (Silva and Bertram, 1991), head-to-medium spacing (Silva and Bertram, 1991; Bertram et al., 1986), polishing induced roughness (Bertram et al., 1986), or general inhomogeneities in the film microstructure (Tarnopolsky et al., 1991). These models are more phenomenological and do not relate directly to the microstructure as do the fundamental calculations of Zhu and Hughes (Sec. 4.3.3). There are enough data showing how micromagnetic modifications through process control have improved noise in thin-film media (Sec. 5.) to discount the notion that fluctuations in macroscopic properties could be solely responsible for thin-film media noise.

4.3.3 Modeling

Important theoretical work has been done in the last decade elucidating the hysteretic and noise performance of thin-film media. Hughes made important theoretical inroads to the modeling of thin film media by calculating hysteresis curves and transition patterns using assumptions mimicking CoP plated media (Hughes, 1983). He used as a base of calculation, a grid of hexagons representing the grains of a magnetic thin film. Each grain is assumed to be a single domain with a uniaxial anisotropy and an assumed easy axis direction. Grain diameter, thickness, and separation could be adjusted. Because of the large channels observed between CoP grains, believed to be nonmagnetic Co_2P, Hughes ignored exchange coupling in his model. Also, because of large demagnetizing fields out of plane, only planar fields were allowed. By increasing the intergranular interactions through reduction of intergranular spacings, he found that he could simulate avalanche reversal (high S^*), the formation of clusters, and the existence of zigzag transition boundaries.

Zhu and Bertram, in a series of important papers, showed that the inclusion of exchange coupling is critical to modeling hysteretic properties and noise properties in all types of thin-film media. (For a complete review of modeling efforts, see Chapter 6). These exchange forces are very short range and tend to align neighboring grains with the same direction of magnetization. Figure 9 shows how the addition of exchange coupling makes the hysteresis loop more square than by use of magnetostatic coupling alone (Zhu and Bertram, 1988). Calculated noise is also much higher in the presence of exchange coupling (Zhu and Bertram, 1988b). Figure 10 compares the mean magnetization and the standard deviation of the magnetization for an exchange coupled array and a non-exchanged coupled array of grains along a recorded transition. The standard deviation is a measure of the magnetization fluctuations across the track and therefore the media noise. The exchange coupled case shows more noise and a localization of the noise in the center of the transition. The Zhu and Bertram efforts have reproduced most key results observed in thin-film media, and have predicted that noise in thin-film media can be reduced by breaking the exchange coupling between grains.

The experimental observation of clusters, the demonstration of the importance of strong magnetostatic and exchange coupling to the hysteretic and noise behavior of thin films, and the experimental characterization of film media noise have defined the direction of low noise thin-film disk fabrication. Films are required with grains that are not exchanged coupled so that magnetic cluster size is minimized.

4.3.4 Measurement of interactions

Minimizing noise in thin-film disks requires understanding of the microstructure and how it affects micromagnetic coupling of grains. Measurements are needed that relate required magnetic properties to the film deposition process variables. Attempts have been made to quantify and correlate more fundamental magnetic and material parameters to the recording noise.

4.3.4.1 Coercive Squareness (S^*)

S^* is easily measured using conventional magnetometers and recently more rapid measuring techniques have become available (Johnson and Kerr, 1990;

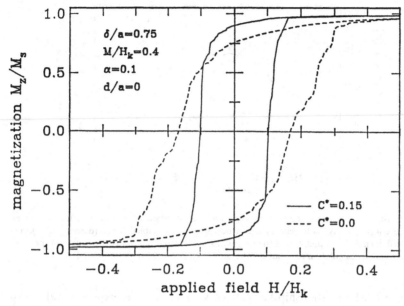

Figure 9. Simulated hysteresis loop with intergranular exchange coupling. C^* is a measure of the exchange energy. C^* is 0.15 for the solid curve. The dashed curve is the result of $C^* = 0$ plotted for comparison. Crystalline anisotropy axes are oriented randomly in three dimensions. The thin-film medium is modeled by a planar hexagonal array of hexagonal shaped grains with film thickness δ, intergranular boundary separation d, grain surface to surface diameter D and lattice spacing $a = d + D$. $H_k = 2K/M$ is the anisotropy field and K the anisotropy energy constant (Zhu and Bertram, 1988) © 1988-IEEE.

Fisher and Pressesky, 1989; Josephs, 1991). Its importance to output signal has been clarified (Williams and Comstock, 1971). In recent years S^* relevance to media noise has been measured and discussed.

Two important studies have recognized that thin-film disk media noise decreases with decreasing values of S^* (Natarajan and Murdock, 1988; Sanders et al., 1989). Figure 11 shows this trend (Sanders et al., 1989). The functionality seems true for a variety of magnetic alloys and different process conditions. The noise increases very steeply in the range of $S^* = 0.85$-0.90, implying the initiation of exchange coupling that results in the observed avalanche switching phenomenon that is linked to increased media noise. It is thus argued that S^* can serve as a measure of the strength of magnetostatic and exchange interactions. Modeling certainly shows that decreased coupling implies lower noise and lower values of S^* (Zhu and Bertram, 1988; Zhu and Bertram, 1988b). A simple approach would be to use the simple parameter S^* as a direct means of characterizing media interactions and noise.

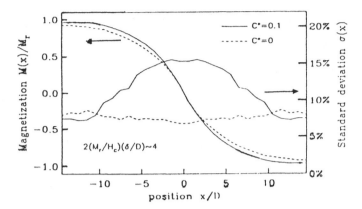

Figure 10. The mean magnetization $\overline{M}(x)$ and the standard deviation $\sigma(x)$ for both a exchange coupled array (solid curve) and a non-exchange coupled array (dashed curve). The standard deviation is a measure of the magnetization fluctuations across the track and provides a direct way to calculate media noise (Zhu and Bertram, 1988b) © 1988-IEEE.

The N_m-S^* relationship has proven very valuable to experimental interpretations, particularly for thin-film disk structures where only small changes in design and process are made. However, all now agree that low noise and low S^* do not always correlate. S^* is not only an indicator of coupling strength, but is also related to the direction of the c-axis of the hcp structures Co alloy grains. As an extreme example, a perpendicular recording film has columnar grains where the easy axis of magnetization is pointing normal to the film plane. S^* measured in the planar direction is extremely small, and yet says nothing about the noise properties of the medium. Degrees of orientation do not have to be so radically different as a perpendicular medium to show S^* effects. Many thin-film disks have been shown to grow epitaxially on the Cr underlayer through the matching of lattice dimensions. One particular match that has often been observed is the (11$\bar{2}$0) plane of Co lying parallel to the disk surface because of a preferred match to the Cr (002) plane below. One can induce the Cr (002) growth by sputtering under high rate and high temperature conditions (Johnson et al., 1990; Duan et al., 1990; Duan et al., 1990b). Films with the c-axis locked in plane invariably show higher S^* values than the other primary Co alloy epitaxy observed where the c-axis is tilted 30 degrees out of plane (Ishikawa et al., 1986; Johnson et al., 1990; Coughlin et al., 1990).

Figure 11 shows a very general trend with observable scatter and and an errant point. Other recent data has shown that there are exceptions to the noise vs. S^* relationship (Speliotis, 1990c; Khan et al., 1990b; Lee et al., 1990).

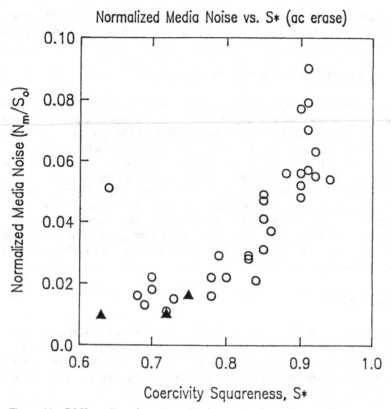

Figure 11. RMS media noise voltage N_m for ac erased media normalized to zero-to-peak isolated pulse amplitude S_0 vs. the coercivity squareness (Sanders et al., 1989).

Speliotis argues that a film can be decoupled having low noise but still have high S^* values from a narrow distribution of anisotropy fields (H_k). This mechanism would explain his results of similar noise values on films with S^* values ranging from 0.70 to 0.93. Khan observed noise to be substantially different between a CoNi and a CoCrTa alloy, even though they both have similar S^* values. The presence of perpendicular anisotropy in the film may be relevant (Sec. 4.3.4.3).

Thin-film media require a mixture of S^* criteria. One wants a reasonably high value of S^* for signal output, writability, and narrow transitions (Yogi et al., 1990b; Speliotis, 1990c; Williams and Comstock, 1971). On the other hand low noise media require uncoupled grains, which is often manifested by a low S^* parameter. The best medium would be one with very high S^*, but uncoupled grains for low noise (Speliotis, 1990c). The S^* parameter would fail to identify such a recording medium. Thus, there has been a search for other measurement techniques that may reveal a more fundamental insight into the media noise without the confounding of anisotropy field distributions and crystallographic axis distributions.

4.3.4.2 Delta-M (δM) Coupling Measurements

It has been recently proposed that Stoner-Wohlfarth (1948) and Henkel (1974) treatments of remanence curves be applied to recording media as a possible means to quantify the effects of interactions simply and quickly (Kelly et al., 1989). Two remanence curves are measured; the isothermal remanent magnetization curve ($I_r(H)$), and the dc demagnetization curve ($I_d(H)$). The I_d curve is measured by initially saturating the sample followed by a sequential measurement of the remanent moment at increasing values of the field in the reverse direction starting at 0 Oe. This process is continued until the sample has reached saturation in the opposite direction. The I_r curve is obtained similarly except that the initial state of the sample is ac demagnetized. Stoner-Wohlfarth theory shows that there is a simple relationship between I_d and I_r for non-interacting particles where $I_r(\infty)$ is the saturation remanence:

$$I_d(H) = I_r(\infty) - 2I_r(H) \tag{3}$$

Henkel proposed that the deviation from this behavior was caused by interactions. By noting the deviation from a straight line in the plot of I_d vs. I_r, a quantification of the interaction could be obtained. An explicit interaction term (δM) as a function of the field H was added (Kelly et al., 1989) to give the following relationship:

$$\frac{I_d(H)}{I_r(\infty)} = 1 - \frac{2I_r(H)}{I_r(\infty)} + \delta M(H) \tag{4}$$

$\delta M(H)$ is twice the difference between the fraction of particles switched at a particular field in the $I_r(H)$ and $I_d(H)$ curves. Figure 12 shows a schematic of an I_r, I_d, and δM curve. This approach gives a new method to explore the media interactions on a more fundamental level and may add information to S^* analyses while avoiding the complications of the effect of other variables.

Kelly's initial work presented δM differences between the circumferential and radial directions in a CoP plated recording media. Several studies have been done since then applying the δM technique to other types of thin-film media. It was demonstrated that for CoNiCr films grown on different Cr underlayer thicknesses, the δM technique correlated with noise (Mayo et al., 1991b). For thin Cr underlayers giving high noise in the magnetic film, the δM measurement

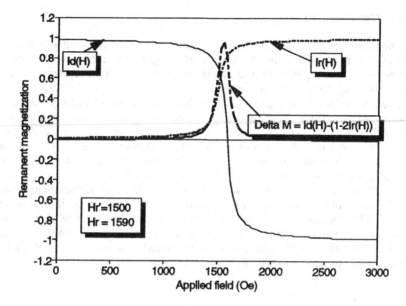

Figure 12. Schematic presentation of a $I_d(H)$, $I_r(H)$ and a δM curve. A Tan⁻¹ function was chosen for the remanence curves. The two curves have identical coercive squareness parameters and only differ in the remanent coercivity values. The $I_r(H)$ and $I_d(H)$ remanent coercivity values were arbitrarily chosen to be $H_{r'} = 1500$ Oe and $H_r = 1590$ Oe respectively. The larger dc remanence coercivity results in positive values of δM and can be explained by the existence of interactions tending to magnetize the sample.

showed strong positive magnetizing interactions. The thick Cr underlayer case (2000 Å) indicated a predominantly negative interaction profile consistent with a demagnetizing interaction analogous to particulate media. Khan et al. (Khan et al., 1991) compared CoCrTa and CoNi thin films using several techniques including δM. The positive δM interaction was 50% higher for the noisier CoNi film, even though the S^* values were 0.89-0.90 for the two films, indicating stronger interactions.

A few theoretical studies have been done on calculating δM curves. Curves were calculated for four different exchange field strengths (Zhu and Bertram, 1991). As the exchange field strength decreases, the positive δM peak decreases and broadens. For zero exchange field, δM becomes mainly negative. This calculation would agree with the CoNiCr work on varied thickness Cr underlayers where the exchange is thought to be changing causing δM and recording noise changes (Mayo et al., 1991b). Beardsley (Beardsley, 1991) performed calculations with added out of plane anisotropy as well as exchange, and found that the curves are affected by both variables. This suggests δM interpretation may suffer from the same ambiguities as S^*. δM measurements and their relationship to thin-film disk noise and structure are still evolving.

4.3.4.3 Perpendicular Anisotropy

The existence of a component of perpendicular anisotropy in longitudinal thin films has prompted some workers to investigate the relationship between such a perpendicular anisotropy and media noise. Yoshida et al. measured CoNiP plated films using torque magnetometry and found that although the easy axis of magnetization lies in the plane of the film, measured anisotropy constants (K_n) were smaller than those of shape anisotropy K_s due to the thin film $(K_s = 2\pi M_s^2)$ (Yoshida et al., 1988). $K_n < K_s$ indicates that perpendicular anisotropy $(K_\perp)$ exists and may play a role in the switching dynamics of thin-film disks. The first noise measurement relative to $K_\perp$ was performed about the same time and noisier disks had smaller $K_\perp$ values (Speliotis, 1988). This noise functionality suggested that three-dimensionally random, or perpendicular preferred alignment of the easy axes of the grains results in lower noise than in-plane alignment. This analysis was extended to several CoNiM and CoCrM ternary alloys where M = Pt, Ta, and Zr (Shiroishi et al., 1988). Media noise was plotted against the total anisotropy constant K_n as measured in a torque magnetometer. The relationship between $K_\perp$ and K_n is the following:

$$K_n = 2\pi M_s^2 - K_\perp \tag{5}$$

where M_s is the saturation magnetization. Figure 13 shows the experimentally determined relationship between noise and K_n. The smaller K_n and larger $K_\perp$ value are, the lower the media noise is. The origin of the perpendicular anisotropy is thought to arise from either the increase in the portion of perpendicularly oriented grains or the reduction of the grain sizes. An argument for decreased noise was based on a reduced demagnetization field mechanism where magnetization at the recorded bit can orient out of plane reducing the transition length and the media noise.

Subsequent work continued to show the media noise - $K_\perp$ relationship (Shiroishi et al., 1989; Speliotis, 1990b). Shiroishi, however, invoked the possibility that the effective noise reduction mechanism may actually be grain segregation, which reduces noise while at the same time induces perpendicular anisotropy. Correlations to $K_\perp$ may be fortuitous. Others compared media noise on CoNi and CoCrTa films and found that the media with the larger $K_\perp$ had lower noise (Khan et al., 1990). Khan, however, concedes that the quieter CoCrTa medium may have both physical and compositional segregation occurring resulting in grain decoupling and quieter films. It was suggested that both mechanisms were possible in a study of noise vs Cr underlayer thickness (Johnson et al., 1990). As the Cr underlayer thickness increased, media noise was reduced. Grains were clearly segregated at the higher thicknesses supporting the decrease

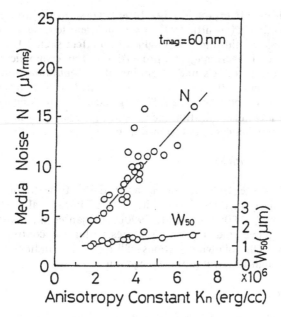

Figure 13. Anisotropy constant dependence of media noise and half pulse width for CoNiM/Cr and CoCrM/Cr (M = Pt, Ta, and Zr) thin-film media with various noise characteristics (Shiroishi et al., 1988) © 1988-IEEE.

in exchange coupling mechanism as being operative. But at the same time, thicker Cr changed its crystallographic state to the (110) configuration promoting Co ($10\bar{1}1$) epitaxy with the c-axis 30 degrees out of plane. This out-of-plane tilt of the c-axis could be seen as support for the mechanism of increasing $K_\perp$ giving lower noise.

Most recent thoughts on the $K_\perp$ mechanism are that it cannot be causal to lower noise levels in thin-film disks. Only a weak dependence of media noise with perpendicular orientation in multilayered CoPtCr/Pd films was found (Sanders et al., 1990). The greatest improvements were achievable through reduced exchange coupling in films with well defined in-plane orientations. Recent evidence was presented on two disks with similar noise levels but very different $K_\perp$ values (Coughlin and Viswanathan, 1991). The noise-$K_\perp$ relationship is apparently not universally correct.

5. Fabrication of Thin-Film Media for Low Noise

The key finding of the last decade is that strong intergranular coupling in in thin-film disks leads to high noise conditions. The exchange interaction is very short range and drops off exponentially. Recent work (Murayama et al., 1991) suggests that separations of as little as 10 Å are sufficient to decouple neighboring grains. It is the goal of thin-film disk engineers to effect such 10 Å separations while maintaining good macromagnetic properties and an economical fabrication process. It is ironic that today's understanding of the optimum thin-film media design is one whose microstructure resembles particulate disk technology that thin-film technology is replacing. The following sections will discuss microstructures necessary for low noise thin-film media and the techniques in general usage to minimize interactions and noise in thin-film media.

5.1 Physical grain segregation

Physical grain segregation has been observed in TEM micrographs in low noise thin-film media (Chen and Yamashita, 1988; Miura et al., 1988; Johnson et al., 1990; Yogi et.al, 1990; Yogi et al., 1990b; Ranjan et al., 1990). There are several means to achieve voided grain structures within the control of the thin-film disk fabricator. The following sections will detail the findings in regards to noise reduction through physical grain segregation.

5.1.1 Process effects

Sputtering process effects on thin-film disk microstructure and media noise have been interpreted using adatom mobility arguments initially put forth by Thornton (Ranjan et al., 1990; Yogi et al., 1990; Thornton, 1986). Thornton envisions film growth in three steps. The first step involves the transport of the atoms to the substrate. The second step involves the adsorption of the atoms onto the surface of the substrate and their diffusion over the surface. The third step relates to bulk diffusion of the atoms to their final position. The transport step is largely determined by sputtering pressure and rate while the diffusion steps are controlled by surface temperature and bias conditions. Thornton has championed the use of structure-zone models to interpret film structure in the framework of these steps. Figure 14 shows a zone diagram for metal films deposited by magnetron sputtering showing the effect of temperature and sputtering pressure on film structure. T is the substrate temperature and T_m is the melting point of the thin-film material. Sputtered films deposited at low temperatures and high pressures fall into Zone 1 and exhibit a columnar growth structure defined by voided open boundaries. Voided grain structures are desired for low noise media. Low sputtering rates, high sputtering pressure, low sputtering temperatures, and the absence of bias are low mobility sputtering conditions that rob the sputtered atoms of their kinetic energy during atom transport and film growth allowing the formation of the voided structures. The following sections will look in more detail at the results of forming magnetic thin films under a variety of low mobility adatom conditions.

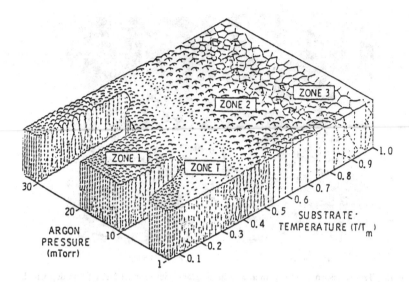

Figure 14. Microstructure zone diagram for metal films deposited by magnetron sputtering. T is the substrate temperature and T_m is the coating material melting point (Thornton, 1986).

5.1.1.1 Pressure and Temperature

Pressure and temperature are two of the important variables used to produce voided grain structures in low noise thin-film media (Yogi et al., 1990; Yogi et al., 1990b). Figure 15 shows the grain structures at low and high temperatures and low and high pressures for CoPtCr/Cr films. The low mobility Zone 1 type disks made at high pressure and room temperature reveal the greater amount of voiding. These disks were quieter and exhibited lower S^* values indicating a reduced amount of exchange coupling. Pressures of 24 mTorr were used to obtain the best results. Similarly, Ar pressures up to 25 mTorr at ambient sputtering temperatures produced the best noise disks in a CoNiCr/Cr system (Ranjan et al., 1990). Sputtered CoNi films on Cr at pressures from 5 to 25 mTorr and temperatures up to 200 °C showed reduced noise (Koga et al., 1989). Low noise results were interpreted in the vein of the low mobility argument, i.e. low kinetic energy conditions lead to films with irregularities on the nm scale.

Many others have observed noise-pressure effects. Chen presented some older work on CoRe revealing marked grain isolation at pressures of 50 and 75 mTorr (Chen and Yamashita, 1988). Noise reduction was observed for a CoCrTa alloy when the sputtering pressure was increased from 0.2 to 10 mTorr (Kawanabe et al., 1990). Prior work by this author, although not providing any noise results, demonstrated that columnar structure appears when sputtering at the higher 10 mTorr pressure. The low mobility of arriving adatoms is invoked

36

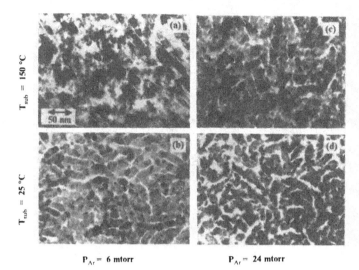

$T_{sub} = 150\,^\circ C$

$T_{sub} = 25\,^\circ C$

50 nm

$P_{Ar} = 6$ mtorr $P_{Ar} = 24$ mtorr

Figure 15. Transmission electron micrographs in plane view for CoNiCr films deposited at $25\,^\circ C$ and $150\,^\circ C$, for 6 and 24 mTorr sputtering pressures. The low mobility magnetic films show a greater amount of voiding (Yogi et al., 1990b) © 1990-IEEE.

as the source of the columnar structure (Kawanabe and Naoe, 1988). Werner et al. noticed the morphology change from high to low pressure on films of CoNi and CoNiCr on Cr underlayers (Werner et al., 1990). Densities were measured on the low and high pressure films and it was found that the density in the high pressure film was 40% lower - an indirect measure of voiding. The Auger depth profiles of the high pressure media also showed an appreciable amount of oxygen in both the Cr underlayer and the magnetic layer. This oxidation of the metal constituents is thought to aid in the morphological separation of crystallites.

Magnetic grain segregated morphology can be induced by sputtering either the underlayer or the magnetic layer at an elevated pressure independent of the sputtering pressure of the other layer (Yogi et al., 1991). Noise reduction follows the reduction in exchange coupling as observed by physical voiding in the microstructure.

The gain in noise reduction from high pressure and low temperature sputtered films does have a negative side - loss in M_s, S, and S^*. A decrease in the values of these variables will have an effect on the Williams-Comstock transition parameter and result in a film with low output at high densities.

5.1.1.2 Bias

Little has been reported of the effect of bias on thin-film media noise. Bias can be considered as one of the adatom mobility variables discussed above (Yogi et al., 1990). The presence of bias means more kinetic energy is available on the surface of the growing films resulting from the bias induced argon ion bombardment onto the substrate. Again, from the mobility argument, the extra available kinetic energy leads to more continuous structures with no granular features. More recent work has shown that biasing only the underlayer or magnetic layer increases noise (Yogi et al., 1991). Bias applied to the underlayer results in a very smooth surface nucleating a continuous film of high noise. Bias applied only to the magnetic layer can counter the effect of a segregated Cr underlayer by giving enough mobility to the adatoms leading to increased coupling and more noise.

An observation counter to this example has been observed (Lu et al., 1990). CoCr/Cr film structures prepared with a -50 V bias added 6 dB to the S/N_m at 1600 fc/mm. Grain size decreased under the bias conditions but no explanation was offered to explain the results as a function of bias voltage.

Bias is a process variable not often utilized in the fabrication of thin-film media. The majority of deposition equipment is based on dc magnetron sputtering; rf sputtering with bias can give the opportunity for microstructure control but the process is very slow compared to dc sputtering. Attention has been focused on other variables besides bias for thin-film microstructure control.

5.1.2 Underlayer effects

The use of underlayers is universal in the fabrication of thin-film media. Their primary purpose has been to nucleate magnetic films with the c-axis in-plane or near in-plane to produce high H_c and high squareness magnetic films. It is now clear that the nature of the underlayer has a large effect in determining noise performance. Magnetic films tend to replicate the underlayer structure below; thus, forming segregated underlayer structures should lead to segregated magnetic film structures. Work by Yogi et al. and Natarajan and Murdock in 1988 revealed the reduction of noise with increasing Cr thickness for sputtered films of CoP and CoNiCr (Yogi et al., 1988; Natarajan and Murdock, 1988). Natarajan and Murdock suggest that columnar growth of the Cr underlayer begins about 1000 Å , the point where a precipitous drop in noise vs. Cr underlayer thickness is observed. Columnar structure could lead to separation of grains through physical voids. Yogi et. al. observed similar results on CoNiCr but with noise reduction beginning at thicknesses as low as 500 Å . Their TEM cross sectional micrographs gave convincing evidence of what Natarajan and Murdock speculated; Cr grains are more granular at increased thicknesses leading to magnetic grain segregation. H_c and S^* increased and decreased respectively with Cr underlayer thickness, a trend in agreement with calculations where decreasing exchange coupling not only decreases noise but increases H_c and drops S^* (Zhu and Bertram, 1988).

Several workers have noted the grain enlarging and segregating effect on Cr microstructure as the Cr thickness is increased. Using high resolution SEM, it was shown that increasing Cr thickness from 400 Å to 3000 Å changed the basic Cr structure from fine grained to acicular Cr grains well separated from each other (Khan et al., 1990). It has been noted that not only does thick Cr give larger and more segregated grains, but that thick Cr does a good job in covering the circumferential texture lines giving an isotropic OR (Coughlin et al., 1990). Signal to media noise vs. increasing Cr thickness for CoPtCr films was studied (Johnson et al., 1990). Underlayers above 500 Å had circular shaped grains about double the diameter of thinner Cr films. In addition, the grains were clearly separated by voids as seen in Fig. 16. Noise dropped, H_c increased, and S^* fell - all indicative of a reduction in exchange coupling as seen in the CoNiCr work by Yogi. Very recent studies of CoNiPt on sputtered Ni-P, have confirmed the very general principle of noise reduction induced by segregated underlayer structures. Ni-P is apparently more effective than Cr for noise reduction because of a larger amount of segregation available with this material (Yamashita et al., 1991).

Other elements have been tried as underlayer materials and the basic principles of grain segregation in the underlayer causing a similar segregation in the magnetic layer hold. The structure of the underlayer is influenced by the process. W and Mo were used as underlayers for CoPtCr films and shown to be noisier at low sputtering pressures but equal at higher pressures (Yogi et al., 1991). Atomic mobility was invoked as the mechanism responsible for the different observed performance. At low pressure, surface kinetic energy of the sputtered atoms is governed by their atomic mass. The heavier atoms have more kinetic energy and this increased mobility leads to more coupled structures and more media noise (Sec. 5.1.1). At the higher pressures, the atomic mass mechanism is supplanted by kinetic energy reduction from increased collision number, and any underlayer leads to magnetic films with similar and low noise results.

Underlayer thickness functionality using W is somewhat different from Cr. Thicker W underlayers have shown an increase in media noise with CoPtCr and CoNiCr magnetic films (Yogi et al., 1989). Noise values were measured to be higher by a factor of ten, compared to the Cr underlayer case, but the result is explainable by increased mobility of the heavier W atom at the lower sputtering pressures. An interesting counterpoint to this work reports that W underlayers reduce media noise by one half at a variety of pressures from 3 to 18 mTorr on CoCr films (Ranjan, 1990). The thickness functionality is the same as Yogi (thicker underlayers of W give higher noise), but the comparison of noise to a

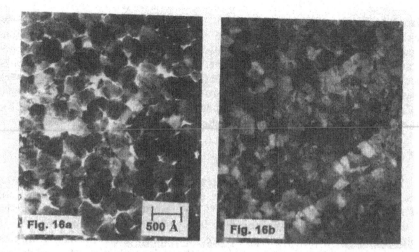

Figure 16. A comparison between grain structures in CoPtCr films sputtered on Cr under-layers of 100Å and 1500Å thickness. The thicker Cr underlayer (Fig. 16a) develops grains with 450Å diameters - about twice that of 100Å Cr underlayer (Fig. 16b). In addition, the thick Cr underlayer reveals a grain structure with physical voiding (Johnson et al., 1990).

conventional CoCr film is in the opposite direction. Subtleties in the sputtering conditions must be invoked to explain the differences between the two investigators.

There are certain drawbacks in using the underlayer thickness method as one's noise reducing strategy. First, thick Cr underlayers require more sputtered material and will decrease disk throughput in a disk manufacturing plant. Second, thicker Cr inevitably increases H_c and decreases S^* through the breaking of exchange coupling. Thus, certain macromagnetic designs may not be attainable with a low noise restriction. Third, lowered S^* values may be indicative of less coupled and quiet films, but low S^* can also have an impact on the recording signal and writability. Furthermore, low S^* values are indicative of hysteresis curves with low S values resulting in low M_r values and decreased high frequency output. Finally, for many cases, the noise sensitivity to Cr thickness is not very strong and only several dB can be gained for rather large Cr thickness increases. Clearly, other noise reduction techniques are desirable for the optimization of macromagnetic and micromagnetic properties.

5.1.3 Magnetic Layer Thickness

Work has been presented that shows thin-film media noise to be a function of the magnetic layer thickness with a steady improvement from 500 to 50 Å (Sanders et al., 1989b) . Cross section TEM micrographs reveal smaller and more isolated grains for the very thin films. It is suggested that as the film grows thicker, the grains coalesce into a more closely packed, more highly exchanged coupled structure exhibiting higher media noise. However, this observation may be more complicated than first thought. Recent work explored the δM curve functionality of different thickness CoPtCr/CrV films down to magnetic thicknesses of 85 Å (Mayo et al., 1991). There is little difference in the δM curves or the first differential of the remanence curves indicating that the coupling nature of the films is independent of the magnetic layer thickness.

5.1.4 Gas Doping

Sputtering Cr underlayer in the presence of N_2 affects the magnetic properties of CoNi films and particularly the media noise (Wakamatsu and Mitobe, 1988). The Cr grains are smallest at 10% incorporation of N_2 into the Cr film. The ensuing replication of the magnetic layer of these small grains leads to an argument that it is the small grain size, or appropriately the larger number volumetric density that is responsible for the noise reduction. Auger electron spectroscopy data does, however, suggest that there may be Cr diffusion into the magnetic layer which could also have a noise reduction effect.

5.2 Compositional Segregation

Compositional segregation of grains in thin-film disks is another effective way to minimize interactions giving low noise disks. Unfortunately, analysis for the presence of compositional segregation is a more difficult measurement than investigating physical grain segregation by spotting intergranular voids using conventional TEM techniques. Therefore, discussion will begin by describing evidence for compositional segregation, followed by subsections on process variables such as substrate temperature, annealing, and magnetic alloy type which can produce low noise thin films through compositional segregation.

5.2.1 Experimental Evidence for Compositional Segregation

Practically all experimental work regarding the segregation of Cr in thin magnetic films has been done on CoCr films used for perpendicular recording. Only very recent work has been done on films suitable for longitudinal recording (Maeda and Takei, 1991; Suzuki et al., 1991). Nevertheless, the work done on CoCr should be indicative of mechanisms in binary and ternary horizontal films.

Macroscopic evidence for composition variations within a thin film came early from the observation of the increase of M_s, the Curie temperature T_c, and H_c compared to the bulk value. (Maeda and Asahi, 1987b). For CoCr films with a Cr content of about 20-22%, the film M_s can increase up to twice that of the bulk M_s (Maeda and Takahashi, 1990). Curie temperatures of CoCr sputtered films change upon heating. A good example is found in $Co_{78}Cr_{22}$ films that showed magnetization up to 700 °C, whereas the bulk alloy, with a T_c of ≈ 300 °C, has lost magnetization at this higher temperature (Snyder et al., 1987). However, if the thin film is heated to 800 °C, an irreversible change occurs - the T_c drops to around 300 °C and the M_s value falls to to 149 emu/cm³, values similar to the bulk CoCr. This data is consistent with explanations involving either Cr grain-boundary segregation or atomic-scale redistribution of the Cr in the original thin film (Snyder et al., 1987). Sputtered CoCr thin films would have Co rich grains necessarily leading to higher M_s values. Extreme heating causes the thin film to recrystallize in the phase of the bulk form.

In addition to M_s increase in CoCr thin films, anomalous increases in H_c have also been observed. As the degree of Cr segregation increases, Hc increases but at a much faster rate than the corresponding increase in the anisotropy constant K (Lodder and Zhang, 1988). Also, the theoretical value of 1 for the demagnetizing factor in perpendicular films is decreased far below 1 for films with Cr segregation (Lodder and Zhang, 1988). In a study on horizontal films, it was found that by varying rf bias, H_c increased in alloys containing Cr and the amount of increase was more rapid with higher Cr content (Tani et al., 1990). It was thus argued that the bias effect is a result of Cr segregation.

Much work has been done attempting to reveal the microstructure responsible for the observed macromagnetic changes. Particularly interesting work has been done on CoCr perpendicular thin films using the technique of selective wet etching (Maeda and Asahi, 1987; Maeda and Asahi, 1987b; Maeda and Takahashi, 1989; Masuya and Awano, 1989; Maeda and Takahashi, 1990). Samples are immersed in a dilute solution of aqua regia, HCl and HNO_3, and observed using TEM. Co regions are preferentially dissolved leaving patterns designated as CP structures (chrysanthemumlike pattern). Structures appear to be more complicated than the conventional models where Cr rich regions occur on grain boundaries (Smits et al., 1984). Figure 17 shows a conventional structure of Cr precipitation to the grain boundaries and a model based on the CP patterns observed from wet etching (Maeda and Asahi, 1987b). Apparently, for these films, the phase separation into ferromagnetic and nonmagnetic phases occurs in each crystallite and phase separation resembles a spinodal decomposition (Maeda and Asahi, 1987). The Co rich stripes orient perpendicularly to grain boundaries suggesting that compositional fluctuations occur parallel to the grain boundaries due to surface or grain boundary diffusion. This observed microstructure suggests that existence of a magnetic unit which is far smaller than the crystallite diameter. Similar but more varied internal grain patterns were observed in a study of the effect of added C to the macromagnetic and micromagnetic properties of CoCr (Masuya and Awano, 1989). More recently, the selective wet etching method has been applied to longitudinal CoPtCr thin films and similar effects have been observed (Suzuki et al., 1990;).

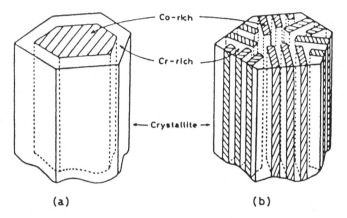

Figure 17. Segregated microstructure models for CoCr perpendicular films. (a) conventional model and (b) CP structure model. The CP structure has been recently observed in longitudinal CoPtCr thin films (Maeda and Asahi, 1987b) © 1987-IEEE.

Attempts at observing composition segregation at the grain level have been tried using other techniques. X-ray microanalysis has been used to measure the local composition at various points within a thin film (Chapman et al., 1986; Rogers et al., 1989). By placing the electron beam in a transmission electron microscope on the grain centers and grain boundaries, Co/Cr ratios can be measured. For films prepared with high substrate temperatures (T_s), clear differences are seen. The high T_s films are found to be markedly inhomogeneous with appreciable composition variations within the column and significant Cr enrichment at the boundaries. Structures within the CoCr columns did not show the CP structures observed by Maeda, but the films were prepared under different conditions. In addition to qualitatively showing composition variations in grains, the use of X-rays allows a quantitative analysis of the elemental segregation. For this set of CoCr films, the Cr content was enriched from 22 atomic percent in the grain to 29 atomic percent at the boundary. X-ray fluorescence microanalysis was also used to investigate CoCr perpendicular films (Maeda and Takahashi, 1989). Results indicate a nonmagnetic Cr rich core to the grain followed by a Co-rich annulus and finally, a Cr enriched grain boundary.

Other more esoteric techniques have lent credence to the hypothesis that Co based magnetic thin films are not compositionally homogenous. As early as 1985, ferromagnetic resonance studies (FMR) showed the presence of several magnetic constituents in a set of CoCr films (Mitchell et al., 1985). Different resonances in the FMR spectrum correspond to different $2K/M_s$ values of the film. It was speculated that one resonance derived from a constituent localized in the interface between the substrate and the CoCr magnetic film. Others have interpreted FMR spectra of CoCr as indicating the formation of two ferromagnetic phases, and that by adjusting substrate temperature prior to film deposition, suitable films with nonmagnetic phases can be prepared (Ma and Schwerdtfeger, 1987). Later work using magnetic resonance showed that indeed, CoCr thin films showed a very different spectrum from the CoCr powder; several resonance lines at higher frequencies indicated the existence of different Co segregation regions (Yoshida et al., 1990; Yoshida et al., 1991). Spectra of CoCr films with either 10% or 25% Cr were almost identical suggesting that the Cr concentration in the ferromagnetic regions did not change; the extra Cr is postulated to go the nonmagnetic grain boundaries. The spin-echo spectra suggests that the ferromagnetic regions have a composition of $Co_{94}Cr_6$ Chemical wet etching confirmed the existence of segregation.

Only recently, have ferromagnetic resonance studies been applied to longitudinal thin films. $CoCr_{12}Ta_2$ was studied using magnetic resonance (Maeda and Takei, 1991). Samples prepared at an elevated temperature indicate the precipitation of a $Co_{95}Cr_5$ alloy. The existence of a Co rich alloy suggests the formation of Cr rich nonmagnetic components. Chemical etching work on the same samples indicates that the segregation occurs both within the grain and at the grain boundaries.

A few other techniques giving indirect evidence of composition segregation have been tried. Mossbauer studies were performed on CoCr films containing a small amount of Fe (Parker, 1986). One of the two features observed in the spectrum was identified as a nonmagnetic phase of nearly pure Cr. Recent Brillouin spin wave scattering studies on CoNiPt films also indicate elemental segregation (Murayama et al., 1990). The distinct and sharp surface acoustic wave spectral features (SAW) coupled with the highly damped standing spin waves (SSW) indicated that Co-rich crystals are surrounded by segregated nonmagnetic alloys of Pt and Ni.

There is plenty of evidence, both from direct measurements of the grain structure and indirect spectral measurements, that vacuum deposited thin films have compositionally segregated structures. Most of the work has been done on CoCr perpendicular films, but very recent work indicates that the general observations are also pertinent to horizontal films. Segregating grains through physical voiding, or creating nonmagnetic regions internal to the grains or on the grain surfaces, should be equally effective in producing decoupled microstructures necessary for low noise performance.

5.2.2 Experimental Methods for Producing Compositional Segregation

Preparation methods to induce compositional segregation seem to be less numerous than physical voiding techniques and less well understood. Compositional segregation has been invoked to explain low noise in longitudinal thin-film media, but only recent analytical work has confirmed that this mechanism is even at work in longitudinal media. Since the observation of the phenomenon of elemental segregation is recent, fabrication methods cannot have been well established. The perpendicular CoCr work clearly shows that sputtering temperature is the most important factor. A few annealing studies show how the effect of post sputtering temperature treatment can aid in elemental segregation. The choice of the alloy composition, particularly in regards to solubility of nonmagnetic constituents, is also key to inducing compositional segregation.

5.2.2.1 Temperature

The recognition of the macromagnetic differences in M_s in deposited CoCr perpendicular thin films compared to the bulk alloys sparked the initial work on composition segregation. Films prepared at different temperatures showed various deviations from the bulk. Figure 18 shows the changes in M_s and H_c along with average Cr content for different substrate temperature values (T_s) (Maeda and Takahashi, 1990). As discussed above, CP structures and Cr rich phases have been invoked to explain these observations. Thus, the judicious selection of sputtering temperature and pre-sputter substrate heating temperatures is a primary method to control the composition segregation in perpendicular films. The effect of substrate temperature on sputtered CoCrTa/Cr longitudinal films has been recently reported (Lu, 1991). Increasing temperature from 25 °C to 220 °C produced a steady decrease in media noise power. Furthermore, δM measurements showed a monotonic decrease in their positive lobe amplitudes indicating a shift from strong exchange forces to the weaker magnetostatic ones. This result probably results from an increase in a Cr rich nonmagnetic phase. Other workers investigating CoCrTa longitudinal films, showed that films deposited at room temperature and 150 °C displayed a migration from a homogenous composition to a segregated one as observed by magnetic resonance techniques (Maeda and Takei, 1991).

Annealing has been tried as a way to alter macromagnetic properties, but it may also have the result of altering micromagnetics properties as well. For a CoNiTa alloy on a Cr underlayer, annealing temperatures above 300 °C increased the H_c value yet dropped the M_s value (Kawanabe et al., 1990b). It was argued that nonmagnetic CoCr constituents were being formed, robbing the magnetic moment. At the same time, Cr from the underlayer was thought to be diffusing into the magnetic layer as observed from the broadening of the interface using Auger spectroscopy. It was proposed that the Cr presence segregated the

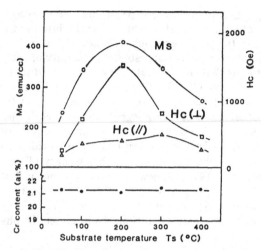

Figure 18. Change in film M_s, H_c, and mean Cr content for CoCr films sputtered with different substrate temperatures (T_s). All the films have approximately the same Cr content. Deviations in M_s and H_c are attributed to the compositional inhomogeneity of the Co and Cr in the grain (Maeda and Takahashi, 1990)

grains producing higher coercivity. Annealing experiments have also been performed on CoNiCr/Cr films (Duan et al., 1990). Microstructures of the magnetic film did not seem to change with annealing even though the H_c, M_s, and S^* values showed appreciable change. The strongest functionality in the macromagnetic property change was the thickness of the Cr underlayer, suggesting that Cr diffused from the underlayer during the anneal to the magnetic film grain boundaries. Although no recording work was done, Lorentz images suggested that the coupling was much reduced on the annealed samples sputtered on thick Cr underlayers. The macromagnetic changes observed by several investigators suggest that annealing is also effective at producing segregated low noise films.

5.2.2.2 Alloy

Coupled very closely with the temperature variable, as a means to initiate compositional segregation, is the selection of the alloy and the understanding of its phase diagram. Choosing an alloy composition that can undergo a phase change or a spinodal decomposition at elevated temperatures is necessary to induce compositional segregation. It is suggested that based on equilibrium phase diagrams, certain binary and ternary alloys can be two phase alloys. Noise performance is better using the alloys that allow two phases such as CoCr and sputtered CoP (Natarajan and Murdock, 1988). Such phase separation may be particularly true for the CoCrTa alloy. At above 2% Ta, the H_c value drops precipitously suggesting that the hcp phase of this composition becomes unstable and that the precipitation of a different phase may be starting (Sellmyer et.al, 1990). The large Ta atom incorporates substitutionally into the Co lattice for low Ta percentages, but as the lattice expands with added Ta, instability is created. The existence of an unstable phase suggests the possibility of the precipitation of nonmagnetic constituents that could isolate the grains. Christner suggests that the good noise behavior of their CoCrTa films may be due to reduced coupling of grains from the inhomogeneous distribution of Cr in the film (Christner et al., 1988). The general observation of CoCr segregation as seen in the CoCr perpendicular films is invoked as a basis for their argument. Compositional segregation may help explain why CoCrTa alloys, often lacking any evidence of a physically voided structure, are less noisy compared to other ternary alloys used for longitudinal recording (Johnson, 1991). A new phase appearance in CoNiTa was used to explain the drop in S^* as compared to CoNi (Khan et al., 1990). CoNi alloys form a complete series of solid solutions, but the added Ta may exceed the solid-solubility limit tending to segregate Ta or some Ta intermetallic phase to the grain boundaries.

Sputtering ternary or even quaternary alloys may prove to be an effective avenue to producing low noise thin-film media because of the possibility of creating nonmagnetic phases within the sputtered film that can segregate grains reducing magnetic couplings. An understanding of the phase diagrams is critical, however, little data exists on phase diagrams on Co based alloys above the binary level. Fabricating new alloys with constituents having limited solubilities may be the key to the next generation of low noise alloys.

5.3 Grain Size

For fixed particle volume packing fraction, media noise power is inversely proportional to the number of magnetic particles per unit volume. This result forms the basis of particulate disk noise reduction strategy (Mee, 1964). In the absence of strong coupling forces in thin-film media, magnetic cluster formation will be suppressed, and small grain size allowing a large grain volumetric density may become active as the noise reduction mechanism. Several workers have reported that for thin films uncoupled in nature, the microstructures having smaller grains have lower noise. Films with similar H_c and S^* macromagnetic parameters showed good correlation between noise and grain size. The lack of change in S^* suggests weak intergranular coupling, opening up the possibility of grain size as a causal factor for the low noise observations. For instance, thin films prepared on thin Cr underlayers that nucleate smaller magnetic grains have shown the effect of decreasing noise as the grain diameters dropped several hundred Å (Mitobe et al., 1988; Moda et al., 1990). Others used thick Cr underlayers to segregate the grains and then controlled size by temperature (Suzuki et al., 1990)). TEM micrographs of these films showed rice-like shaped grains well separated and apparently uncoupled from each other. The recording results show noise dependence proportional to the square root of grain size.

The use of high Ar sputtering pressure along with N_2 gas doping has the effect of forming a finer grain structure, and a noise reduction has been observed (Wakamatsu and Mitobe, 1988). Again, these workers observed no change in macromagnetics suggesting that coupling is weak and constant. This would allow grain size effects to come to the forefront. Shiroishi et al., as discussed earlier in Sec. 4.3.4.3, correlated a noise reduction with increasing perpendicular component measured in the film (Shiroishi et al., 1989). But it seems that along with this observation, grain size also decreased. If grain coupling is reduced, other film properties such as perpendicular component and grain size take control over the noise properties of the film. All sputtering process variables affect grain size to some degree, and if one can produce uncoupled films by adjusting one variable, and reduce magnetic grain size by using another, low noise films can be fabricated.

5.4 Substrate Effects

The substrate surface effect on both macromagnetic and micromagnetic properties has been considered for thin-film media. Most commercial film media today have a circumferential texture imparted to the Ni-P after a preliminary polish operation. Texturing reduces the contact area between head and disk reducing stiction and friction concerns. The roughness of the polish is about 20 Å RMS, while the roughness of the texture varies from 50-100 Å RMS. It has been observed by many workers that the circumferential texture gives rise to a circumferential anisotropy in the macromagnetic properties (Teng and Ballard, 1986; Simpson et al., 1987; Lin et al., 1989; Johnson, 1991). Both M_r and H_c are typically higher in the texture direction. The effect on noise has also been reported on but with less clear results (Judge and Speliotis, 1987).

Experiments have been described to increase orientation ratio (OR) by circumferentially texturing substrates (Mirzamaani et al., 1990). It seems that OR appears if the texture is circumferential and the Cr underlayer has crystallized with $<100>$ preferred orientation, allowing the CoPtCr to grow epitaxially with a $<11\bar{2}0>$ preferred orientation. These preferred orientations are a function of sputtering conditions, particularly the preheating temperatures that remove water and contaminants from the substrate surface. However, TEM micrographs of thin-film disk surfaces in disks with $OR > 1$ did not show any evidence of the effect of texture, such as grain structure chained along the grooves, leaving the origin of the anisotropy somewhat a mystery.

Others have argued for OR originating from an orientation of crystalline axes along the texture lines through grain boundary matching when the grain diameters and the substrate roughness are of the same physical scale (Lin et al., 1989). However, their electron diffraction patterns do not give any direct evidence of anything other than random planar distributions of the Co (10$\bar{1}$1) planes. It may be that the scratches are directly causing the anisotropy (Coughlin et al., 1990). With a thin Cr underlayer, the texture is visible using high resolution SEM; in thicker Cr films the grooves are masked. The mechanism for OR is the breaking of intergranular coupling in the radial direction, resulting in a higher OR. δM curves measured perpendicular and parallel to the circumferential grooves on a plated CoP thin-film disk, also suggest that interactions of the grains are different for the two directions (Kelly et al., 1989).

Stress has been suggested as a variable responsible for OR. Data has recently been presented where the grains seemingly do line up with texture grooves (Kawamoto and Hikami, 1991). Rather than assuming intergranular coupling of the grains in the circumferential direction as the source of OR, the X-ray diffraction data implies that the lattice is compressively strained along the texture grooves. Inverse magnetostriction can now be invoked as the source of anisotropy. Earlier work with different Pt concentrations in the magnetic film showed a correlation between OR and Pt composition (Nishikawa et al., 1989). This correlation is similar to the one between magnetostriction and Pt content reported earlier by Aboaf (Aboaf et al., 1983). Recent work by Doerner et al. presents some circumstantial evidence for magnetostriction; since X-ray diffraction implies that the Cr underlayer is under tensile stress in the direction normal to the film plane, it must be under compressive stress in the tangential direction (Doerner et al., 1991). This stress is speculated to be responsible for OR. The amount of stress is controllable by sputtering temperature, sputtering pressure, and substrate texturing and surface chemistry.

Whatever the actual mechanism for OR, it is clear that microstructural changes are inducing macromagnetic changes resulting in the anisotropy. The question arises whether the couplings, crystallographic orientations, or strains on the grains are having an effect on media noise. Signal to noise ratios up to 2000 fc/mm with disks of increasing OR were studied with no apparent effect. A disk with OR < 1 did have poorer S/N_m performance, but it is rooted in low signal from low M_r and S^* values affecting the transition parameter. A similar experiment extended the recording densities to 4000 fc/mm (Doerner et al., 1991). Now, the OR $\sim$ 1 disks outperformed the higher OR disks in S/N_m, although it is not clear whether this improved performance is from a drop in noise or increase in signal. A similar result was reported that with increasing H_c orientation ratio, media noise increased (Koga et al., 1989). It was proposed that the zigzag transitions would become more distorted when having to cross the irregularities of texture line barriers that produced the higher OR.

Pure roughness effects of the substrate have also been considered to be degrading factors on the media noise. Fluctuations in fly height induced by texture have been suggested to be a significant source of media noise (Bertram et al., 1986). These claims were countered by comparing recording properties of CoNiCr/Cr magnetic film structures sputtered on mirror-polished Ni-P surfaces to films sputtered on a circumferentially textured Ni-P surface (Seagle et al., 1987). The textured disks were superior in S/N_m, signal, and noise separately. The improved signal was attributed to increased squareness and coercivity parallel to the circumferential texture. The lowering of the media noise with texture was not explained. Judge and Speliotis studied electrolessly plated CoP on textured and non-textured surfaces (Judge and Speliotis, 1987). As Seagle et al. observed, the signal improves because of larger squareness and much decreased switching field distributions in the recording direction. However, media noise is primarily independent of texturing in their work for thicknesses of magnetic media from 600-800 Å. The smoother surfaces they studied showed only an incremental noise improvement at 250 Å magnetic thickness.

Others have correlated noise increases to surface parameters more subtle than roughness. Koga et al. presented data where noise in a CoNiCr thin film on a circumferentially textured Ni-P substrate increased when RMS values were increased from 20 to 100 Å (Koga et al., 1989). But when films were sputtered on glass substrates with increasing roughnesses, the noise was constant. Koga postulated that some sort of surface irregularity other than roughness in the textured Ni-P was responsible for the introduction of a noise-creating mechanism in the CoNiCr. The notion that noise is not a function of roughness alone was further corroborated by comparing noise performance on two disks, one having been ion milled prior to sputtering. Assuming no change in surface roughness, the drop in noise with ion milling had to be attributed to some other surface irregularity. Similarly, Katti and Saunders saw no correlation between surface roughness and media noise using the dc erase noise technique (Katti and Saunders, 1990). Media noise had no dependence on surface roughnesses between 10 Å and 50 Å. They argue that fluctuations in some unspecified media parameters, and not the roughness, are responsible for any observed increase in media noise.

These results seem counterintuitive to the conventional wisdom requiring ultra smooth surfaces for high density recording. One might imagine that microscopic irregularities may cause noise problems in thin-film media. Glass companies are betting on the need of the rigid disk industry for optically smooth glass substrates. But it may be that ultra smooth surfaces facilitate very uniform grain growth and increased grain coupling, cancelling any beneficial effect of a smooth surface. Smooth surfaces certainly do allow for closer flying of recording heads without incurring collisions with surface asperities. Perhaps the lower fly height allowed is the primary benefit of low roughness surfaces.

5.5 Multilayers

There has been much interest recently in the use of magnetic multilayers as a means to decrease noise in thin-film media. Interest has been kindled in multilayers (sometimes termed compositionally modulated films) as a new class of materials with novel properties. (Chang and Giesen, 1985; Shinjo and Takada, 1987). A number of systems have been studied to understand the effects of the interface and reduced dimensionality on magnetic properties. In thin-film magnetic recording media, many benefits also abound from the use of multilayers. As with other techniques and variables that affect thin film magnetic properties, multilayers were first used to gain control over the macromagnetic properties. It was found that the coercivity could be adjusted independently from the $M_r\delta$ product by making structures with one to fifteen layers of Co separated by Cr (Lazzari et al., 1967; Maloney, 1979; Maloney, 1971). By decoupling the Co alloy magnetic layers, one can obtain the moment of the sum of all the layers, but the H_c of only one layer. Later work on CoNiCr and CoNi alloys showed similar relationships (Katayama et al., 1988; Tokushige and Miyagawa, 1990). It is an empirical fact that H_c drops with increasing layer thickness for thin film disk structures of any alloy, process, or underlayer combination. Since most workers want to maintain a high H_c, the individual layer thicknesses of the Co magnetic alloy are adjusted to obtain the desired coercivity while maintaining magnetic moment by controlling the number of layers. This method also saves one from depositing extremely thick Cr underlayers, which are effective in obtaining high coercivities but consume significant amounts of time and material. (Katayama et al., 1988).

Although multilayers have been shown to have a potential benefit for H_c and $M_r\delta$ control, macromagnetic control of thin-film recording disks has been accomplished by other means such as alloy composition, underlayer effects, and sputtering process modifications. Multilayering was not considered as a viable method for macromagnetic control primarily because of the extra process complications of adding other layers, and because other more effective methods were available. Other drawbacks of multilayers are the increased effective thickness causing a to increase and writability signal amplitude to decrease. More recently, however, it has been shown that multilayering can have important effects on media noise and that the extra process complications may be worth the trouble.

Figure 19 shows a schematic cross section of a typical bilayer disk with a Cr interlayer. Figure 20 shows an Auger depth profile of a bilayer produced in our laboratory. The Cr curve clearly shows the presence of a 25 Å interlayer sandwiched between two equally thick magnetic layers. Hata et al. were the first to report improved noise performance (Hata et al., 1990). They produced a double layer CoNiCr thin film, each layer of 250 Å thickness, separated by a Cr interlayer with varying thicknesses up to 500 Å. Recording signal was slightly

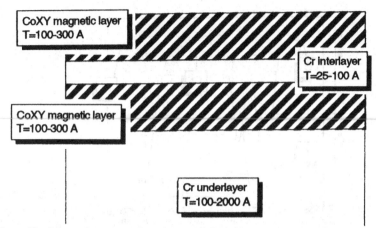

Figure 19. Schematic cross section of a multilayered thin-film disk with two magnetic layers separated by a Cr interlayer. Cr interlayers less than 50Å have proven most effective.

degraded but assignable to an increase in spacing loss; on the other hand, a reduction in media noise was observed. They attributed the lowered noise to a decoupling of the two CoNiCr layers by the Cr interlayer. For equally thick uncoupled magnetic layers, the total noise power adds as the sum of the squares of the individual layers noise voltages; the net result is that the multilayered total film RMS noise voltage drops as the square root of the number of layers. Hata cites further evidence for the decoupling of the two CoNiCr layers by the reduction of the coercive squareness, S^*, as predicted by calculations (Zhu and Bertram, 1988).

The noise equation for a two layer disk can be presented as the following, where N_T is the total noise voltage for the entire structure and N_1 and N_2 are the individual noise contributions from the separate layers (Murdock et al., 1990).

$$N_T^2 = N_1^2 + N_2^2 + 2 < N_1 N_2 > \tag{6}$$

The magnitude of the noise in each individual layer and the disposition of the cross term dictates the total noise quantity. Fabricating the multilayered structure with intrinsically quiet single layers is the first step to ensuring a quiet multilayered media structure. Choosing magnetic alloys that are quiet because of exchange decoupling through composition modulation, or selecting very thin layer building blocks whose grains are physically segregated is a natural way to initiate a multilayered structure. Elemental segregation has already been discussed in Section 4.3. The use of thin layers in multilayers has been proposed (Lambert et al., 1990). Because of the lack of grain coalescence in the early development of the

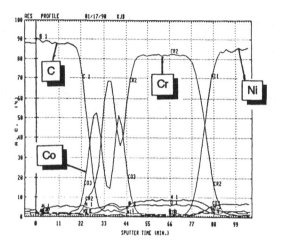

Figure 20. Auger depth profile of a bilayer disk. The Cr curve clearly shows the presence of a 25Å interlayer sandwiched between two equally thick magnetic layers.

film structure, these films would naturally be decoupled. Evidence for the existence of decoupled and quiet films as a function of thickness for very low thicknesses has been presented (Sanders et al., 1989b). The suggestion of segregated grains for very thin films was supported by later TEM evidence observed in a quadruple layered structure as compared to the single layer control (Hata et al., 1990b). Judicious choices for the multilayer film building blocks can greatly decrease the noise of the final structure.

But more important than the low noise layer constituents, is the nature of the cross term in Eq. 6. For a fully coupled structure, exchange coupling will dominate and align the moments parallel in the two layers. The cross term will be positive and equal to $2N^2$ and no noise reduction will be realized. If exchange and magnetostatic coupling is eliminated, the random nature of the interactions will cause the cross term to go to zero because the noise sources would be statistically uncorrelated. Noise will be reduced because the number of grains per unit volume is increased. If the Cr interlayer is thick enough to break off the exponentially decaying exchange forces, yet small enough to allow for significant magnetostatic interactions, the cross correlation term can be negative thus resulting in even more noise reduction.

Several studies have emerged on the effect of interlayer thickness on the noise of multilayered disks. Interlayer thickness is one way to adjust the magni-

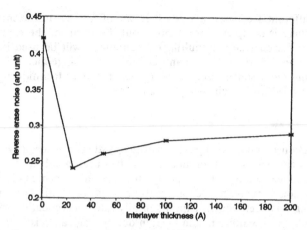

Figure 21. Reverse erase noise for a series of bilayer CoPtCr magnetic films separated by varying thicknesses of Cr. The noise minimum occurs at 25Å. The asymptotic noise value approaches $1/\sqrt{2}$ of the noise with no interlayer. This agrees with a model where exchange and magnetostatic interactions are equal to zero.

tude of the magnetostatic interaction. Murdock observed that a Cr interlayer of 50-100 Å gave the greatest noise reduction (Murdock et al., 1990). Noise reduction was still significant out to 1000 Å of Cr interlayer, suggesting that long range magnetostatic interactions were present. Others have observed the same long range interaction but with noise minima occurring at Cr interlayer thicknesses as low as 25 Å (Min et al., 1991; Johnson, 1991b; Palmer et al., 1991). An accompanying minimum in S^* was also observed indicating maximum decoupling. Figure 21 shows the noise as measured by the reverse dc erase technique for a series of bilayer films made with CoPtCr separated by varying Cr thicknesses (Johnson, 1991b). S^* in the figure parallels the noise reduction. δM measurements in a similar study, show a significant drop in the positive peak indicating a decreased interaction as the Cr interlayer thickness goes from 300 to 25 Å (Min et al., 1991). It has been suggested that stress may be playing a role in the the improved recording properties of multilayered thin-film disks (Ranjan et al., 1991). X-ray diffraction spectra of bilayer structures with thick Cr interlayers show shifted and broadened Co peaks, suggestive of a higher stress component and decreased in-plane crystalline anisotropy in the layered structures.

Conventional signal related recording measurements of multilayered films show the expected results from macromagnetic parameter changes of multilayers as compared to single layer films (Palmer et al., 1991; Miller et al., 1991). The most interesting and by far the most dominating observation is the improvement in S/N_m, peak jitter, and on track error rate. Reduction in media noise by using multilayered techniques has been confirmed in at least six separate laboratories using several different Co based alloys - it is one of the unusual scientific observations where there is complete agreement about the result of the experiment. It still remains to be seen whether multilayer technology will be used in thin-film disk manufacturing. There are still many effective ways to reduce thin-film disk media noise using more simple and economical single layer techniques and those will probably be used until their utility is exhausted.

6.Summary

The reduction of noise in digital recording systems is necessary to achieve large recording densities. Thin-film media noise has been a substantial contributor to the total noise picture. Fortunately, progress has been made in understanding media noise origins and creating techniques for media noise reduction. Grain decoupling has been shown experimentally and theoretically as the mechanism responsible for low noise media. Grain decoupling can arise from physical grain segregation or compositional segregation. Decoupled microstructures are manipulated through process variable adjustment. Thin-film media noise improvements has put media noise on a par with head and electronics noise contributions.

7.References

Aboaf, J.A., S. R. Herd, E. Klokholm, "Magnetic Properties and Structure of Cobalt-Platinum Thin Films", *IEEE Trans. Magn.*, **MAG-19,** 1514 (1983).

Aoi, H., M. Saitoh, N. Nishiyama, R. Tsuchiya, and T. Tamura, "Noise Characteristics in Longitudinal Thin-Film Media", *IEEE Trans. Magn.*, **MAG-22,** 895 (1986).

Arnoldussen, T.C., E.M. Rossi, A. Ting, A. Brunsch, J. Schneider, and G. Trippel, "Obliquely Evaporated Iron-Cobalt and Iron-Cobalt-Chromium Thin Film Recording Media", *IEEE Trans. Magn.*, **MAG-20,** 821 (1984).

Arnoldussen, T.C., and H.C. Tong, "Zigzag Transition Profiles, Noise, and Correlation Statistics in Highly Oriented Longitudinal Film Media", *IEEE Trans. Magn.*, **MAG-22,** 889 (1986).

Aspland, M., G.A. Jones, and B.K. Middleton, "Properties of Electroless Cobalt Films", *IEEE Trans. Magn.*, **MAG-5,** 314 (1969).

Barany, A.M., and H. N. Bertram, "Transition Noise Model for Longitudinal Thin-Film Media", *IEEE Trans. Magn.*, **MAG-23,** 1776 (1987).

Bate, G., "Recent Developments in Magnetic Recording Materials", *J. Appl. Phys.*, **52,** 2447 (1981).

Baugh, R.A., E.S. Murdock, and B.R. Natarajan, "Measurement of Noise in Magnetic Media", *IEEE Trans. Magn.*, **MAG-19,** 1722 (1983).

Beardsley, I.A., "Significance of δ M Measurements in Thin-Film Media", *IEEE Trans. Magn.*, **27,** 5037 (1991).

Belk, N.R., P.K. George, and G.S. Mowry, "Noise in High performance Thin-Film Longitudinal magnetic Recording Media", *IEEE Trans. Magn.*, **MAG-21**, 1350 (1985).

Bertram, H.N., K. Hallamasek, and M. Madrid, "DC Modulation Noise in Thin Metallic Media and Its Application for Head Efficiency Measurements", *IEEE Trans. Magn.*, **MAG-22**, 247 (1986).

Bonn, T.H. and D.C. Wendell, Jr., *U.S. Patent No. 2,644,787* (1953).

Brenner, A. and G.E. Riddell, *J. Res. Nat. Bur. Stand.*, **37**, 31 (1946).

Chang, L.L., and B.C. Giessen (eds.), Synthetic Modulated Structures, Academic Press, Orlando, FL. (1985).

Chapman, J.N., I.R. McFadyen, and J.P.C. Bernards, "Investigation of Cr Segregation Within RF-Sputtered CoCr Films", *J. Magn. Magn. Mater.*, **62**, 359 (1986).

Chen, T., "The Micromagnetic Properties of High Coercivity Metallic Thin Films and Their Effects on the Limit of Packing Density in Digital Recording", *IEEE Trans. Magn.*, **MAG-17**, 1181 (1981).

Chen, T., and G. B. Charlan, "High Coercivity and High Hysteresis Loop Squareness of Sputtered Co-Re Thin Film", *J. Appl. Phys.*, **50**, 4285 (1979).

Chen, T., D.A. Rogowski, and R.M. White, "Microstructure and Magnetic Properties of Electroless Co-P Thin Films Grown on an Aluminum Base Disk Substrate", *J. Appl. Phys.*, **49**, 1816 (1978).

Chen, T., and T. Yamashita, "Physical Origin of Limits in the Performance of Thin-Film Longitudinal Recording Media", *IEEE Trans. Magn.*, **24**, 2700 (1988).

Christner, J.A., R. Ranjan, R. L. Peterson, and J. I. Lee, "Low-noise Metal Medium for High-Density Longitudinal Recording", *J. Appl. Phys.*, **63**, 3260 (1988).

Coughlin, T., J. Pressesky, S. Lee, N. Heiman, and R.D. Fisher, "Effects of Cr Underlayer Thickness and Texture on Magnetic Characteristics of CoCrTa Media", *J. Appl. Phys.*, **67**, 4689 (1990).

Curland, N., and D.E. Speliotis, "Transition Region in Recorded Magnetization Patterns", *J. Appl. Phys.*, **41**, 1099 (1970).

Daval, J., and D Randet, "Electron Microscopy on High Coercive Force CoCr Composite Films", *IEEE Trans. Magn.*, **MAG-6**, 768 (1970).

Doerner, M.F., P.W. Wang, S.M. Mirzamaani, D.S. Parker, and A.C. Wall, "Chromium Underlayer Effects in Longitudinal Magnetic Recording", *Materials Research Society Symposium Proceedings, Magnetic Materials: Microstructure and Properties*, **232**, (1991)

Dressler, D.D., and J.H. Judy, "A Study of Digitally Recorded Transitions in Thin Magnetic Films", *IEEE Trans. Magn.*, **10**, 674 (1974).

Duan, S.L., J.O. Artman, K. Honom and D.E. Laughlin, "Improvement of the magnetic Properties of CoNiCr Thin Films by Annealing", *J. Appl. Phys.*, **67**, 4704 (1991).

Duan, S.L., J.O. Artman, B. Wong, and D.E. Laughlin, "The Dependence of the Microstructure and Magnetic Properties of CoNiCr/Cr Thin Films on the Substrate Temperature", *IEEE Trans. Magn.*, **26**, 1857 (1990).

Duan, S.L., J.O. Artman, B. Wong, and D.E. Laughlin, "Study of the Growth Characteristics of Sputtered Cr thin Films", *J. Appl. Phys.*, **67**, 4913 (1990).

Eltoukhy, A.H., "A Review of Thin Film Media for Magnetic Recording", *J. Vac. Sci. Technol.*, **A4**, 539 (1986).

Fisher, R.D., and J.L. Pressesky, "Spatially Resolved, In-Situ Measurements of the Coercive Force of Thin Film Magnetic Media", *IEEE Trans. Magn.*, **25**, 3414 (1989).

Hata, H., T. Fukuichi, K. Yabushita, M. Umesaki, and H. Shibata, "Low Noise Media Using Double-Layer CoNiCr Thin Films for Longitudinal Recording", *J. Appl. Phys.*, **67**, 4692 (1990).

Hata, H., T. Hyohno, T. Fukuichi, K. Yabushita, M. Umesaki, and H. Shibata, "Magnetic and Recording Characteristics of Multilayer CoNiCr Thin Film Media", *IEEE Trans. Magn.*, **26**, 2709 (1990).

Henkel, *Phys. Stat. Sol.*, **7**, 919 (1974).

Howard, J.K.,"Thin Films for Magnetic Recording Technology: a Review", *J. Vac. Sci. Technol.*, **A4**, 1 (1986).

Hughes, G.F., "Magnetization Reversal in Cobalt-Phosphorus Films", *J. Appl. Phys.*, **54**, 5306 (1983).

Ishikawa, M., N. Tani, T. Yamada, Y. Ota, K. Nakamura, A. Itoh, "Film Structure and Magnetic Properties of CoNiCr/Cr Sputtered Thin Film", *IEEE Trans. Magn.*, **MAG-22**, 573 (1986).

Johnson, K.E., "Thin-Film Recording Media: Challenges for Physics and Magnetism in the 1990s", *J. Appl. Phys.*, **69**, 4932 (1991).

Johnson, K.E.,"Multilayered Thin-Film Recording Media", CMRR Workshop, Micromagnetic and Microstructural Characterization of Media and Heads for Magnetic Recording, (1991).

Johnson, K.E., P.R. Ivett, D.R. Timmons, M. Mirzamaani, S.E. Lambert, and T. Yogi, "The Effect of Cr Underlayer Thickness on Magnetic and Structural Properties of CoPtCr Thin Films", *J. Appl. Phys.*, **67**, 4686 (1990).

Johnson, K.E., and M.G. Kerr, "A Magnetometer for the Rapid and Nondestructive Measurement of Magnetic Properties on Rigid Disk Media", *IEEE Trans. Magn.*, **26**, 2565 (1990).

Josephs, R.M., "The Remanent Moment Magnetometer - A New Instrument for Non-Destructive Characterization of the Hysteretic Properties of Thin-Film Recording Media" *5th Joint MMM-Intermag Conference*, CE-14 (1991).

Judge, J.S., and D.E. Speliotis, "The Effect of Texturizing on the Magnetic and Recording Properties of Plated Rigid Disks", *IEEE Trans. Magn.*, **MAG-23**, 3402 (1987).

Katayama, S., T. Tsuno, K. Enjoji, N. Ishii, and K. Sono, "Magnetic Properties and Read Write Characteristics of Multilayer Films on a Glass Substrate", *IEEE Trans. Magn.*, **24**, 2982 (1989).

Katti, R.R., and D.A. Saunders, "Thickness and Roughness Dependence of DC Modulation Noise in Thin Film Magnetic Recording Disk Media", *IEEE Trans. Magn.*, **26,** 2712 (1990).

Kawamoto, A., and F. Hikami, "Magnetic Anisotropy of Sputtered Media Induced by Textured Substrate", *J. Appl. Phys.*, **69,** 5151 (1991).

Kawanabe, T., K. Hasegawa, and M. Naoe, "Magnetic and Recording Characteristics of CoCrTa/Cr Thin film Media Prepared by FT Sputtering", *IEEE Trans. Magn.*, **26,** 1593 (1990).

Kawanabe, T., K. Hasegawa, S. Ono, S. Nakagawa, and. M. Naoe, "Cr Migration in CoNiTa/Cr Films by Annealing", *IEEE Trans. Magn.*, **26,** 42 (1990).

Kawanabe, T., and M. Naoe, "Effects of Ta Addition in CoNi/Cr Double Layer Film Sputtered in Low Ar Gas Pressure", *IEEE Trans. Magn.*, **24,** 2721 (1988).

Kelly, P.E., K.O'Grady, P.I. Mayo, and R.W. Chantrell, "Switching Mechanisms in Cobalt-Phosphorous Thin Films", *IEEE Trans. Magn.*, **25,** 3381 (1989).

Khan, M.R., R.D. Fisher, and N. Heiman, "DC-Magnetron Sputtered CoNiTa for High Density Longitudinal Recording", *IEEE Trans. Magn.*, **26,** 118 (1990).

Khan, M.R., S.Y. Lee, S.L. Duan, J.L. Pressesky, N. Heiman, D.E. Speliotis, and M.R. Scheinfein, "Correlations of Modulation Noise with Magnetic Microstructure and intergranular Interactions for CoCrTa and CoNi Thin-Film Media", *J. Appl. Phys.*, **69,** 4745 (1991).

Khan, M.R., S.Y. Lee, J.L. Pressesky, D. Williams, S.L. Duan, R.D. Fisher, N. Heiman, M.R. Scheinfein, J. Unguris, D.T. Pierce, R.J. Celotta, and D.E. Speliotis, "Correlations of Magnetic Microstructure and Anisotropy with Noise Spectra for CoNi and CoCrTa Thin Film Media", *IEEE Trans. Magn.*, **26** 2715 (1990).

Kitada, M., S. Asada, and N. Shimizu, "Magnetic Properties of Sputtered Co-Ir Thin Films", *Thin Solid Films*, **113,** 199 (1984).

Koga, N., S. Ito, and H. Tomiyasu, "Media Noise of Sputtered Metallic Thin Film Disks", *IEEE Trans. Magn. Japan*, **4,** 680 (1989).

Kogure, T., S. Katayama, and N. Ishii, "High Coercivity Magnetic Hard Disks Using Glass Substrates", *J. Appl. Phys.*, **67,** 4701 (1991).

Lambert, S.E., J.K. Howard, and I.L. Sanders, "Reduction of Media Noise in Thin Film Media by Lamination", *IEEE Trans. Magn.*, **26,** 2706 (1990).

Lazzari, J.P., I. Melnick, and D. Randet, "Thin Evaporated Films with High Coercive Force", *IEEE Trans. Magn.*, **MAG-3,** 205 (1967).

Lee, J.I., and P.K. George, "Demagnetization-Free Longitudinal Recording on Flexible Thin-Film Metal Media", *IEEE Trans. Magn.*, **MAG-21,** 1221 (1985).

Lee, S.Y., J.L. Pressesky, D. Williams, and N. Heiman, "Write Current Dependence of Transition Noise in Thin-Film Media", *IEEE Trans. Magn.*, **26,** 121 (1990).

Lin, T., Alani, and D.N. Lambeth, "Effects of Underlayer and Substrate Texture on Magnetic Properties and Microstructure of a Recording Medium", *J. Magn. Magn. Mater.*, **78**, 213 (1989).

Lodder, J.C., and L.C. Zhang, "The Influence of the Segregated Microstructure on the Magnetization Reversal in CoCr Films", *J. Magn. Magn. Mater.*, **74**, 74 (1988).

Lu, M., "Effects of Substrate Temperature on Intergranular Interactions and Recording Noise of CoCrTa/Cr Thin Film Media", *IBM Shared University Research Contract, "High Coercivity Low Noise Thin-Film Media"* (1991).

Lu, M., J.H. Judy, and J.M. Sivertsen, "Effects of RF Bias on the Texture, Magnetics, and Recording Properties of RF Sputtered CoCr/Cr Longitudinal Thin Film Media", *IEEE Trans. Magn.*, **26**, 1581 (1990).

Luborsky, F.E., "Development of Elongated Particle Magnets", *J. Appl. Phys.*, **32**, 171S (1961).

Ma, C.L., and C.F. Schwerdtfeger, "Ferromagnetic Resonance Studies of DC Magnetron Sputtered CoCr Films", *Solid State Communications*, **64**, 651 (1987).

Madrid, M., and R. Wood, "Transition Noise in Thin Film Media", *IEEE Trans. Magn.*, **MAG-22**, 889 (1986).

Maeda, Y., and M. Asahi, "Segregated Microstructure in Sputtered CoCr Film revealed by Selective Wet Etching", *J. Appl. Phys.*, **61**, 1972 (1987).

Maeda, Y., and M. Asahi, "Segregation in Sputtered Co-Cr Films", *IEEE Trans. Magn.*, **MAG-23**, 2061 (1987).

Maeda, Y., and M. Takahashi, "Thermomagnetic Analysis of Compositional Separation in Sputtered CoCr Films", *J. Appl. PHys.*, **68**, 4751 (1990).

Maeda, Y., and M. Takahashi, "Direct Observation of the Segregated Microstructures within Co-Cr Film Grains", *Jap. J. Appl. Phys.*, **28**, L248 (1989).

Maeda, Y., and K. Takei, "Compositional Inhomogeneities in CoCrTa/Cr Films for Longitudinal Recording", *IEEE Trans. Magn.*, **27**, 4721 (1991).

Maloney, W.T., "The Optimization of Sputtered Co-Cr Layered Medium For Maximum Areal Density", *IEEE Trans. Magn.*, **MAG-17**, 3196 (1981).

Maloney, W.T., "Sputtered Multilayer Films for Digital Magnetic Recording" *IEEE Trans. Magn.*, **MAG-15**, 1135 (1979).

Masuya, H., and H. Awano, "Segregated Microstructure and Crystal Structure in Sputter Deposited CoCr Carbon Added Films", *Jap. J. Appl. Phys.*, **28**, 372 (1989).

Mayo, P.I., K. O'Grady, R.W. Chantrell, J.A. Cambridge, I.L. Sanders, T. Yogi, and J.K. Howard, "Magnetic Measurement of Interaction Effects in CoNiCr and CoPtCr Thin Film Media", *J. Magn. Magn. Mater.*, **95**, 109 (1991).

Mayo, P.I., K. O'Grady, P.E. Kelly, J. Cambridge, I.L. Sanders, T. Yogi, and R.W. Chantrell, "A Magnetic Evaluation of Interaction and Noise Characteristics of CoNiCr Thin Films", *J. Appl. Phys.*, **69**, 4733 (1991).

Mee, C.D., *"The Physics of Magnetic Recording"*, North-Holland, Amsterdam, 1964.

Middleton, B.K., "The Recording and Reproducing Processes", *Magnetic Recording Volume 1: Technology,* eds., C.D. Mee and E.D. Daniel, McGraw-Hill, New York, 62 (1987).

Middleton, B.K., and J.J. Miles, "Sawtooth Magnetization Transitions and the Digital Recording Properties of Thin Film Recording Media", *Eighth International Conference on Video, Audio, and Data Recording* **(Conf. Publ. No. 319),** 2 (1990).

Miller, M.S., P.K. George, T.A. Madsen, E.M. Simpson, and J.P. Walber, "Optimization of Low Noise Media for $100 \, Mb/in^2$ Recording with a Magnetoresistive Head", *J. Appl. Phys.,* **69,** 4715 (1991).

Min, T., J.G. Zhu, and J.H. Judy, "Effects of Inter-layer Magnetic Interactions in Multilayered CoCrTa/Cr Thin Film Media", *IEEE Trans. Magn.,* **27,** 5058 (1991).

Mirzamaani, M., K.E. Johnson, D. Edmonson, P. Ivett, and M. Russak, "Orientation Ratio of Sputtered Thin-Film Disks", *J. Appl. Phys.,* **67,** 4695 (1990).

Mirzamaani, M., M. Re, S.E. Lambert, A. Praino, T.S. Petersen, and K.E. Johnson, "Signal to Noise Ratio of Thin-Film Disks with Various Orientation Ratios", *IEEE Trans. Magn.,* **26,** 2457 (1990).

Mitobe, Y., H. Wakamatsu, A. Kakehi, and M. Shinohara, "Investigation of S/N Property on Sputtered Metal Films", *IEEE Trans. Magn. Japan,* **3,** 562 (1988).

Mitchell, P.V., A. Layadi, N.S. VanderVen, and J.O. Artman, "Direct Observation of Magnetically Distinct Regions in CoCr Perpendicular Recording Media Using Ferromagnetic Resonance", *J. Appl. Phys.,* **57,** 3976 (1985).

Miura, S., T. Yamashita, G. Ching, and T. Chen, "Noise and Bit Jitter Performance of CoNiPt Thin Film Longitudinal Recording Media and its Effect on Recording Performance", *IEEE Trans. Magn.,* **24,** 2718 (1988).

Murayama, A., M. Miyamura, and Y. Oka, "Interlayer Exchange Coupling in Co/Cr/Co Double-Layered Recording films Studied by Spin-Wave Brillouin Scattering", *IEEE Trans. Magn.,* **27,** 5064 (1991).

Murayama, A., Murayama, M. Miyamura, S. Ishikawa, and Y. Oka, "Brillouin Study of Spin Waves in Sputtered CoNiPt Alloy Films", *J. Appl. Phys.,* **67,** 410 (1990).

Murdock, E.S., B.R. Natarajan, and R.G. Walmsley, "Noise Properties of Multilayered Co-Alloy Magnetic Recording Media", *IEEE Trans. Magn.,* **26,** 2700 (1990).

Natarajan, B.R., and E.S. Murdock, "Magnetic and Recording Properties of Sputtered Co-P/Cr Thin- Film Media", *IEEE Trans. Magn.,* **24,** 2724 (1988).

Nishikawa, R., T. Hikosaka, K. Igarashi and M. Kanamaru, "Texture-Induced Magnetic Anisotropy of CoPt Films", *IEEE Trans. Magn.,* **25,** 3890 (1989).

Noda, K., Y. Notohara, N. Koga, and H. Tomiyasu, "A Study of the Thinner Cr Underlayer in Co-Alloy/Cr for Rigid Disk", *IEEE Trans. J. Magn. Japan,* **5,** 843 (1990).

Palmer, D., K.E. Johnson, E. Wu, and J. Peske, "Recording Properties of Multi-layered Thin-Film Media", *IEEE Trans. Magn., 27,* 5307 (1991).

Parker, F.T., "Mossbauer Effect of Dilute Fe in sputtered CoCr Films", *J. Appl. Phys., 60,* 2498 (1986).

Ranjan, R., "Beta Tungsten Underlayer for Low-Noise Thin-Film Longitudinal Media", *J. Appl. Phys., 67,* 4698 (1990).

Ranjan, R., J.A. Christner, and D.P. Ravipati, "Effect of Grain Isolation on Media Noise in Thin-Film Longitudinal Media" *IEEE Trans. Magn., 26,* 322 (1990).

Ranjan, R., M.S. Miller, and P.K. George, "Magnetic, Recording, and Crystalline Properties of Multilayered Longitudinal Thin-Film Media", *J. Appl. Phys., 69,* 4727 (1991).

Rogers, D.J., J.N. Chapman, J.P.C. Bernards, and S.B. Luitjens, "Determination of Local Composition in CoCr Films Deposited at Different Substrate Temperatures", *IEEE Trans. Magn., 25,* 4180 (1989).

Sanders, I.L., J.K. Howard, S.E. Lambert, and T. Yogi, "Influence of Coercivity Squareness on Media Noise in Thin-Film Recording Media", *J. Appl. Phys., 65,* 1234 (1989).

Sanders, I.L., D.R. Wilhoit, S.E. Lambert, G.L. Gorman, T. Yogi, and V.S. Speriosu, "Media Noise in Periodic Multilayered Magnetic Films with Perpendicular Anisotropy", *J. Appl. Phys., 68,* 1791 (1990).

Sanders, I.L., T. Yogi, J.K. Howard, S.E. Lambert, G.L. Gorman, and C. Hwang, "Magnetic and Recording Characteristics of Very Thin Metal-Film Media", *IEEE Trans. Magn., 25,* 3869 (1989).

Seagle, D.J., N. C. Fernelis, and M. R. Khan, "Influence of Substrate Texture on Recording Parameters for CoNiCr Rigid Disk Media" *J. Appl. Phys., 61,* 4025 (1987).

Sellmyer, D.J., D. Wang, and J.A. Christner, "Magnetic and Structural Properties of CoCrTa Filmas and Multilayers with Cr", *J. Appl. Phys., 67,* 4710 (1990).

Shinjo, T., and T. Takada, *Metallic Superlattices, Elsevier Science Publishers, New York,* 213 (1987).

Shiroishi, Y., Y. Matsuda, S. Hishiyama, H. Suzuki, T. Ohno, and Y. Yohisa, "Read and Write Characteristics of CoNiZrM/Cr Thin Films for Longitudinal Recording", *IEEE Trans. Magn., 25,* 3390 (1989).

Shiroishi, Y., Y. Matsuda, K. Yoshida, H. Suzuki, T. Ohno, N. Tsumita, M. Ohura, and M. Hayashi, "Read and Write Characteristics of Co-Alloy/Cr Thin Films for Longitudinal Recording", *IEEE Trans. Magn., 24,* 2730 (1988).

Silva, T.J., and H.N. Bertram, "Magnetization Fluctuations in Uniformly Magnetized Thin-Film Recording Media", *IEEE Trans. Magn., 26,* 3129 (1990).

Simpson, E.M., P.B. Narayan, G.T.K. Swami, and J.L.Chao, "Effect of Circumferential Texture on the Properties of Thin- Film Rigid Recording Disks", *IEEE Trans. Magn.,* **MAG-23,** 3405 (1987).

Smits, J.W., S.B. Juitjens, and F.J.A. den Boreder, "Evidence for Microstructural Inhomogeneity in Sputtered CoCr Thin Films", *J. Appl. Phys.*, **55**, 2260 (1984).

Snyder, J.E., K.R. Mountfield, and M.H. Kryder, "Thermomagnetic Analysis, Annealing Effects, and Cr Distribution in CoCr Sputtered Films", *J. Appl. Phys.*, **61**, 3146 (1987).

Speliotis, D.E., "Correlation Between Modulation Noise and Uniaxial Anisotropy in High Coercivity Thin Film Recording Media", *IEEE Trans. Magn.*, **26**, 2721 (1990).

Speliotis, D.E., "Peak Shift in Particulate and Thin-Film Media", *J. Magn. Magn. Mater.*, **83**, 455 (1990).

Speliotis, D.E., "Correlation of Rotational Magnetic Properties and Noise in Thin-Film Magnetic Recording Media", *IEEE Trans. Magn.*, **24**, 2979 (1988).

Stoner, E.C., and E.P. Wohlfarth, "A Mechanism of Magnetic Hysteresis in Heterogeneous Alloys", *Philos. Trans. R. Soc. London,* **240 A 826,** 599 (1948).

Suzuki, H., N. Goda, S. Kojima, S. Nagaike, Y. Shiroishi, N. Shige, and N. Tsumita, "Compositional Separation of CoCrPt/Cr Films for Longitudinal Recording and CoCr/Ti Films for Perpendicular Recording", *IEEE Trans. Magn.*, **27**, 4718 (1991).

Suzuki, H., N. Tsumita, M. Hayashi, Y. Shiroishi, and Y. Matsuda, "Magnetic and Crystallographic Properties of Sputtered CoNiZr Films on Cr, Mo, and W Underlayers", *IEEE Trans. Magn.*, **26**, 2280 (1990).

Talke, F.E., and R.C. Tseng, "An Experimental Investigation of the Effect of Medium Thickness and Transducer Spacing on the Read-Back Signal in Magnetic Recording Systems", *IEEE Trans. Magn.*, **MAG-9,** 133 (1973).

Tanaka, H., H. Goto, N. Shiota, and M. Yanagisawa, "Noise Characteristics in Plated Co-Ni-P Film for High Density Recording Medium", *J. Appl. Phys.*, **53,** 2576 (1982).

Tani, N., M. Hashimoto, M. Ishikawa, Y. Ota, K. Nakamura, and A. Itoh, "Increase of Coercive Force in Sputtered Hard Disk", *IEEE Trans. Magn.*, **26,** 1282 (1990).

Tarnopolsky, G.J., H.N. Bertram, and L.T. Tran, "Magnetization Fluctuations and Characteristic Lengths for Sputtered CoP/Cr Thin-Film Media", *J. Appl. Phys.*, **69,** 4730 (1991).

Tarnopolsky, G.J., L.T. Tran, A.M. Barany, H.N. Bertram, and D.R. Bloomquist, "DC Modulation Noise and Demagnetizing Field in Thin Metallic Media" *IEEE Trans. Magn.*, **25,** 3160 (1989).

Teng, E., and N. Ballard, "Anisotropy Induced Signal Waveform Modulation of DC Magnetron Sputtered Thin Film Disks", *IEEE Trans. Magn.*, **MAG-22,** 579 (1986).

Thornton, J.A., "The Microstructure of Sputter-Deposited Coatings." *J. Vac. Sci. Technol.*, **A4,** 3059 (1986).

Tokushige, H., and T. Miyagawa, "Magnetic Properties of Sputtered CoNiCr/Cr Multilayer Films", *IEEE Trans. J. Magn. Japan,* **5,** 575 (1990).

Tong, H.C., R.Ferrier, P. Chang, J. Tzeng, and K.L. Parker, "The Micromagnetics of Thin-Film Disk Recording Tracks" *IEEE Trans. Magn.*, **MAG-20**, 1831 (1984).

Velu, E.M.T., and D.N. Lambeth, "CoSm Based High Coercivity Thin films For Longitudinal Recording" *J. Appl. Phys.*, **69**, 5175 (1991).

Wakamatsu, H., and Y. Mitobe, "Study on Underlayer and S/N Property of CoNi Disk" *IEEE Trans. Magn. Japan*, **3**, 666 (1988).

Werner, A., H. Hibst, and H. Mannsperger, "Influence of Ar Pressure on Morphology and Recording Characteristics of Hybrid Sputtered Magnetic Disks", *IEEE Trans. Magn*, **26**, 115 (1990).

Williams, M.L., and R.L. Comstock, "An Analytical Model of the Write Process in Digital Magnetic Recording", *AIP Conf. Proc. Magn. Magn. Mater.*, **5**, 738 (1971).

Yamashita, T., L.H. Chan, T. Fujiwara, and T. Chen, "Sputtered Ni_xP Underlayer for CoPt Based Thin Film Magnetic Media" *IEEE Trans. Magn.*, **27**, 4727 (1991).

Yamashita, T., G.L. Chen, J. Shir, and T. Chen, "Sputtered ZrO2 Overcoat with Superior Corrosion Protection and Mechanical Performance in Thin-Film Rigid Disk Application", *IEEE Trans. Magn.*, **24**, 2629 (1988).

Yogi, T., G. Castillo, G.L. Gorman, M.A. Kakalec, and T. Nguyen, "Transition Noise CoPtCr and CoNiCr Media with W Underlayer", *J. Magn. Soc, Japan, Supplement, No. S1*, **13**, 30P (1989).

Yogi, T., G.L. Gorman, C.H. Hwang, M.A. Kakalec, and S.E. Lambert, "Dependence of Magnetics, Microstructures, and Recording Properties on Underlayer Thickness in CoNiCr/Cr Media", *IEEE Trans. Magn.*, **MAG-24**, 2727 (1988).

Yogi, T., T.A. Nguyen, S.E. Lambert, G.L. Gorman, and G. Castillo, "Microstructure and Recording Noise of Thin-Film Longitudinal Media", *Mat. Res. Soc. Symp. Proc.*, **232**, 3 (1991).

Yogi, T., T.A. Nguyen, S.E. Lambert, G.L. Gorman, and G. Castillo, "Role of Atomic Mobility in the Transition Noise of Longitudinal Media", *IEEE Trans. Magn.*, **26**, 1578 (1990).

Yogi, T., T. Nguyen, S.E. Lambert, G.L. Gorman, M.A. Kakalec, and G. Castillo, "Role of Atomic Mass of Underlayer Material in the Transition Noise of Longitudinal Media", *J. Appl. Phys.*, **69**, 4749 (1991).

Yogi, T., C. Tsang, T.A. Nguyen, K. Ju, G.L. Gorman, and G. Castillo, "Longitudinal Media for 1 Gb/in² Areal Density", *IEEE Trans. Magn.*, **26**, 2271 (1990).

Yoshida, K., H. Kakibayashi, and H. Yasuoka, "Magnetic and Microstructural Studies on CoCr Films Investigated Using NMR", *Mat. Res. Soc. Symp. Proc.*, **23**, 47 (1991).

Yoshida, K., H. Kakibayashi, and H. Yasuoka, "Study of CoCr Films for Perpendicular Magnetic Recording Using Nuclear Magnetic Resonance", *J. Appl. Phys.*, **68**, 705 (1990).

Yoshida, K., T. Okuwaki, N. Osakabe, H. Tanabe, Y. Horiuchi, T. Matsuda, K. Shinagawa, A. Tonomura, and H. Fujiwara, "Observation of Recorded Magnetization Patterns by Electron Holography", *IEEE Trans. Magn.*, **MAG-19,** 1600 (1983).

Yoshida, K., T. Yamashita, and M. Saito, "Magnetic Properties and Media Noise of CoNiP Plated Disks", *J. Appl. Phys.,* **64,** 270 (1988).

Zhu, J.G., and H.N. Bertram, "Micromagnetic Studies of Thin Metallic Films", *J. Appl. PHys.,* **63,** 3248 (1988).

Zhu, J.G., and H.N. Bertram, "Recording and Transition Noise Simulations in Thin-Film Media", *IEEE Trans. Magn.,* **24,** 2706 (1988).

Zhu, J.G., and H.N. Bertram, "Self Organized Behavior in Thin-Film Recording Media", *J. Appl. Phys.,* **69,** 4709 (1991).

CHAPTER 3

MEASURED NOISE IN THIN FILM MEDIA

EDWARD S. MURDOCK
Hewlett-Packard Laboratories
Palo Alto, California 94304

1. Introduction

All signals in the real world exhibit some kind of noise; magnetic recording signals are, of course, no exception. The challenge is to characterize, understand and learn to control the noise in order to recover desired signals. All parts of a magnetic recording system contribute to the total noise: the recording head, the electronics of the channel, and, of course, the recording medium itself. Various physical and materials features of the magnetic recording layer create irregularities in the pattern of magnetization that translate into noise on the signal when it is read by the recording head.

All areas of science and experience encounter random fluctuations. Voltage across a resistor and current from a current source fluctuate, as do the amount of rainfall at a given location, the density and speed of traffic on a highway (especially during commute hours!), and the rate of one's heart beat. Characterizing or measuring these fluctuations can be a complex task that depends considerably on the reason the measurement is needed. For instance, resistors exhibit the well-known Johnson noise, electronic fluctuations that translate into fluctuations in the voltage drop across a resistor. If one desires the overall effect of this noise on a circuit, it is sufficient to characterize Johnson noise by a single number, the rms voltage of the fluctuations over the entire bandwidth of interest. On the other hand, if the detailed effect of Johnson noise on a circuit is needed then it becomes necessary to know the rms fluctuation voltage as a function of the frequency of the fluctuations - a set of numbers. In addition, even more precise measurements and analysis are needed if one's goal is to understand the physical cause of Johnson noise (or of any other kind of noise). It is the same for magnetic recording media. Depending on the needs, experimenters employ various measurement techniques. Each kind of measurement gives some information about the media of interest, but the relation between different types of measurement is not always evident.

Magnetic recording media noise is of interest for two reasons. First, it affects the reliability of the data channel. Therefore it is quite important to measure the medium noise, understand its characteristics and develop models for predicting its effect on the channel (in particular, the error rate). The second reason for measuring medium noise is that it is useful for understanding some properties of the recording medium itself and of the recording process that encodes data as a magnetization pattern in that medium. For instance, the existence and importance of quantum mechanical exchange coupling between the grains in thin metal film media was uncovered through characterization of medium noise. In any discussion of medium noise, therefore, it is important to keep in mind which of these two purposes is in view. The kind of instruments used, the techniques employed and even the units of measurement all depend on the purpose of the measurements.

In this chapter we will examine the characteristics of medium noise as it is experimentally observed. Following a discussion of the measurement instruments and techniques commonly used we will discuss several types of medium noise, that is, medium noise observed for several different recorded conditions or remanent states of the medium. The emphasis will be on the observed properties of medium noise with some discussion of the physical causes of these properties.

2. History of Medium Noise Measurements

Noise in the recording channel causes uncertainty in the time at which the readback pulses are detected, thus introducing a probability of error in detection. In general, the physical causes of noise on the readback signal are statistically independent of one another. For instance, fluctuations in the magnetization pattern of the recorded medium are not causally related to the Johnson noise of the recording head. As a consequence, the various noise sources can be studied independently of one another, with the aim of understanding of the causes and control of the magnitude of the noise. Mathematically the *noise* in a recording channel can be represented by the *variance* in the detected location of the recorded transitions (in time or space). Then the total noise is the result of the statistically independent individual contributing noise sources as the equation below shows:

$$\sigma_1^2 + \sigma_2^2 + \ldots + \sigma_n^2 = \sigma_{tot}^2 \qquad (1)$$

where the subscripts refer to the several noise sources.

Before the introduction of thin metal film media in the early 1980's the medium noise component in recording systems was measured by "dc-erasing" a single track on the disk and measuring the contribution of the magnetic medium with a power meter or spectrum analyzer. The non-magnetic noise contribution is easily removed, as will be discussed shortly (in Section 3.1.1), and the resulting "noise voltage" was typically combined with the readback amplitude of isolated pulses to calculate the signal to noise ratio (SNR). This approach worked well for many years for predicting the error rate performance of particulate media. However, as it happened, this method was inadequate for characterizing the new metal film media, whose noise characteristics were unexpected and very different. For a number of years from about 1978-1983 every paper defined and measured medium noise *differently* (Ogawa and Ogawa, 1979; Fisher, et. al., 1981; Tanaka, et. al., 1982; and Terada, et. al., 1983). As a result it was difficult to compare the noise performance of different manufacturers' media or to predict anything about the actual contribution to error rate by any given type of media.

Additionally, because the detailed properties of medium noise for the new thin metal film media were quite different from that of particulate media (especially the frequency spectrum), there was considerable confusion and disagreement about what the observed phenomena meant. For instance, several workers noticed that the total amount of medium noise increased significantly as the linear density of the recorded signal increased (Tanaka, et. al., 1982 and Terada, et. al., 1983). Baugh, Murdock and Natarajan (Baugh, et. al., 1983) clearly showed that the "noise" measured in thin film media is almost exclusively associated spatially with the transitions. In particulate media, the fluctuations in magnetization are associated with discrete particles and clusters of particles and end up uniformly distributed throughout the film. In contrast, the noise in thin metal film media -- that is, the fluctuations in magnetization vector of the track -- is concentrated in certain repeated locations along the track with the regions in between being very quiet, nearly free of fluctuations. This meant, for instance, that the traditional goal of defining a signal-to-noise ratio (SNR) for such media had to be done very carefully. The SNR, to be useful for comparing different media and for predicting the media contribution to error rate, must be measured quite differently than for particulate media. In fact, SNR is not a particularly useful figure of merit for characterizing thin film medium noise.

3. Measurement Methods

Later in this volume there will be a discussion of practical techniques to measure medium noise. Nonetheless, in order to properly understand the measured properties of medium noise it is necessary to say a few words about how it is measured.

3.1 Instruments

3.1.1 Spectrum analyzers

For the purpose of studying noise in recording systems the spectrum analyzer is clearly the primary instrument of choice. It measures the frequency content of the noise voltage and thereby provides a tool useful for gaining insight into the causes of the various noise processes, as well as for predicting some of the channel effects of medium noise. Figure 1 shows typical noise voltage spectra measured on one disk at two different recording densities, 3.75 and 15 MHz. The lowest curve shown (c) is the noise spectrum of the head and electronics alone, measured with the disk stopped so there is no contribution from the recording medium.

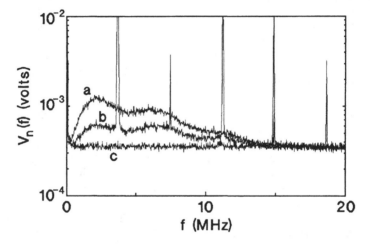

Figure 1:   Noise voltage spectra measured on one disk using a thin film hea at two different recording frequencies. (a) Recorded at 15 MH (60 kfci); (b) Recorded at 3.75 MHz (15 kfci); (c) Background noise of head + electronics, measured with disk stopped.

Much of the spectrum analyzer's usefulness is because the reciprocity relation in magnetic recording between the magnetization pattern on the disk and the magnetic field of the sensing head is a correlation integral as seen in Eq. 2 (Bracewell, 1978, pp. 25 and 46).[1] As such it is equally well represented in either the

---

[1] Most discussions of the reciprocity relation assume that the integral is a convolution integral. Strictly speaking, however, as Bracewell makes clear, an equation of this sort is a cross-correlation integral because of where the minus sign falls in the argument of the magnetization term in the integrand. If the reciprocity equation were a convolution integral its integrand would have the form: $\{h(x)m(vt-x)\}$. However, when calculating $|V(f)|^2$, or power, there is in practice no difference since all quantities in the argument are real functions.

time domain (as an oscilloscope measures) or the frequency domain (as a spectrum analyzer measures). However, the terms of the integral are more easily separable experimentally in the frequency domain. Mathematically,

$$V(t) = c \int dz \int dy \left| dx \ h_x(x,y) \ \frac{d}{dt} m_x(x\text{-}vt,y) \right.$$

(2)

or, performing the Fourier transform,

$$V(f_x) = c' 2\pi \, f_x \, H_x(f_x) M_x(f_x)$$

(3)

where $V(f_x)$ is the Fourier transform of $V(t)$; c and c′ are constants necessary to make the equation quantitative and to relate $V(t)$ to such things as the number of turns, head efficiency and disk/head velocity; $h_x(x,y)$ is the head field per unit mmf; $m_x(x\text{-}vt)$ is the media's magnetization pattern; and f_x is the spatial frequency in cycles/meter. (Spatial frequency is independent of disk/head speed and is the preferred unit of measure because we are dealing with a pattern imprinted in space along a circumferential track. Spatial frequency, f_x, is related to the direct spectrum analyzer unit of temporal frequency -- Herz or cycles per second -- by the head/disk relative velocity, viz., $f_x = f/v$.) In Eq. 3, $M_x(f_x)$ is the Fourier transform of the complete x-component of magnetization, including both noise and signal. Because noise and signal add, this term is a sum of the signal and magnetization noise transforms. The noise term can be considered separately from the signal. Also, since the intrinsic noise of the medium adds incoherently across the track, the track width dependence of noise voltage differs from that of signal voltage and is shown explicitly in Eq. 4:

$$V_n(f_x) = (\mu_o N \varepsilon v \delta) w^{1/2} \, 2\pi f_x H_x(f_x) M_n(f_x)$$

(4)

Because the frequency domain form of the reciprocity relation is a simple product it is clearly easier to deal with experimentally than the integral form in Eq. 2. The key is to carefully note what a spectrum analyzer actually measures and then to derive or measure the transfer function of the head, $H_x(f_x)$.

A prerequisite to effectively reducing medium noise is an understanding of its causes. This requires careful characterization of medium noise, and one essential feature is to measure the intrinsic magnetization fluctuations that cause noise in the signal. Now, a spectrum analyzer measures the time-averaged signal power in a narrow bandwidth centered on a frequency that is swept over the range of interest. Because it measures power it loses all information on the phase of the signal; because it measures a time average of a moving surface it averages the noise over a long stretch of the recorded track with many transitions. This actually simplifies the experimental task, though it does sacrifice some information about the magnetization pattern on the disk; the exact spatial details of the magnetization fluctuation pattern are lost. However, the statistical properties of the noisy magnetization are obtainable. Mathematically, the noise *power* spectral density which the spectrum analyzer measures is:

$$S_n(f_x) = (\mu_o N \varepsilon v \delta)^2 \, w \left| 2\pi f_x H_x(f_x) \right|^2 S_\phi(f_x) \qquad (5)$$

where $S_n(f_x)$ is the measured power per unit bandwidth in $\text{Volt}^2/(\text{cycle/meter})$, as a function of spatial frequency; $|H_x(f_x)|^2$ is the squared magnitude of the head field transfer function, obtained analytically, by numerical modeling, or by Fourier transform of a measured head field; $S_\phi(f_x)$ is the noise power spectral density of the magnetization fluctuations along the track; the factor $(2\pi f_x)^2$ comes from differentiating the magnetization pattern in the reciprocity relation; and w is the track width. (Some spectrum analyzers can be set to measure noise voltage spectral density, $\sqrt{S_n(f)}$, directly; otherwise it must be calculated from the measured voltage spectrum by normalizing by the resolution bandwidth. A Hewlett-Packard Application Note tells how to do it correctly (Anonymous HP Application note, 1976). (It is not correct to simply divide by the square root of the Resolution Bandwidth, when measuring noise.) Also, NOTE that the *spatial* frequency, f_x (in cycles/meter), is substituted here for the temporal frequency, f (in cycles/second), which is the actual ordinate which the spectrum analyzer measures. Spatial frequency is the natural unit, as mentioned above. Some care must be taken in converting the measured noise voltage spectrum in volts at each temporal frequency, $V_n(f)$, to noise voltage spectral density as a function of spatial frequency, $\sqrt{S_n(f_x)}$).

In quantitative studies of medium noise it is necessary to express the noise power spectral density (NPSD) using the correct dimensions. This is particularly true when trying to understand the causes of medium noise or to calculate the spatial extent of noise correlations, both of which require analysis of the intrinsic noise power spectral density of magnetization fluctuations, $S_\phi(f_x)$. Expressing the noise power spectral density with the correct dimensions is not necessarily straightforward, since the NPSD is a quantity related to fluctuations in magnetization and spectrum analyzers directly measure *volts*. Converting the spectrum analyzer's measurement of the raw noise spectrum into the measurement of noise power spectral *density* has units of $S_n(f) = \text{Volt}^2/\text{Hz} = \text{Volt}^2 \cdot \text{sec}$. Changing the independent variable to spatial frequency, f_x, is accomplished by noting that $f_x = f/v$ (v = head/disk velocity), so that *measured*

$$S_n(f_x) = S_n(f) \cdot v = \frac{Volt^2}{(cycle/meter)} = Volt^2 \cdot meter \qquad (6)$$

From this it can be shown that the dimensions of $S_\phi(f_x)$ in Eq. 5 are *Amp²* (in SI units).

Now, $S_\phi(f_x)$ is the term which contains the physics of the medium noise; that is, it alone in Eq. 6 contains the information about the physical fluctuations in magnetization in the medium. The other terms in the equation are the head transfer function and scaling constants, none of which contain any information about the magnetization pattern. As mentioned above, S_ϕ is the noise power spectral density (NPSD) of the micromagnetic fluctuations in the magnetization along the track (actually, because of the reciprocity relation, S_ϕ is the NPSD of the x-component of magnetization fluctuations, averaged across the track and through the depth of the medium sampled by the head). In order to properly interpret the measured spectrum of the noise it is essential that $S_\phi(f_x)$ be computed from the measured noise spectrum, which is done according to Eq. 5. Only by removing the filtering effect of the head on the spectrum can the magnetization spectrum be clearly seen and the physical phenomena that cause medium noise be understood. Examples will be found in later sections in this chapter.

Briefly, however, under some recorded conditions the noise has $S_\phi \propto 1/f_x^\alpha$ while other conditions give $S_\phi \propto e^{-2\pi l x f x}$ (Baugh, et. al., 1983). These different statistics indicate that different physical processes are at work generating the magnetization fluctuations. These details are masked if the raw noise power spectrum is observed, because the head transfer function dominates the shape of the raw noise spectra, making the differences due to different recorded conditions not readily apparent. Nonetheless, once $|2\pi f_x H_x(f_x)|^2$ has been removed it is instantly obvious that the two proportionalities mentioned are different and point to different physical causes of the noise. (See Fig. 3 & 4.) It is important to note, however, that division of $S_n(f_x)$ by $(2\pi f_x)^2$ as required by Eq. 5 amplifies small errors at low frequencies, requiring very careful measurements at low frequencies to obtain accurate results.

For many purposes, including comparing media and predicting overall effect of noise on error rate, the broadband noise power is a useful measure. $V_n(f)$, the measured noise voltage spectrum, is commonly used to compute the total broadband noise power for a given recording medium, giving a result that is similar to a measurement with an rms voltmeter and representing the noise power which will combine with the noise from the channel electronics and the recording head to determine the error rate. Referring to Fig. 1, the integrated medium noise power, P_n, is given by the area between the spectrum of the electronics background noise and the spectrum of the combination of (medium noise + electronics noise), namely,

$$P_n = \int_0^{f_{max}} \left[S_{n,total}(f) - S_{n,elec}(f) \right] df \qquad (7)$$

where f is the frequency (in Hertz), $S_{n\ total}(f)$ and $S_{n\ elec}(f)$ are the measured *power* spectral densities (in volts^2/Hz), and the integration is over the desired bandwidth up to f_{max}. $S_{n\ total}(f)$ is computed from the raw measured spectrum, $V_n(f)$, by squaring V_n and normalizing correctly by the resolution bandwidth of the spectrum analyzer to obtain a spectral density (HP Application Note #207, 1976). The upper integration limit, f_{max}, is determined by the purpose of the measurement. For fundamental studies one needs to include all medium noise power, so $f_{max} = \infty$, effectively. On the other hand, for practical predictions of noise impact on a given recording channel, f_{max} would be the upper corner frequency of the input low pass filter of the channel.

3.1.2 Rms voltmeters

A spectrum analyzer, while useful as an instrument for obtaining details of the noise which are necessary for understanding and controlling it, is rather slow. If a fast measurement is needed and simply the total medium noise power in a given bandwidth is required, an rms voltmeter with an appropriate filter is a suitable instrument.

3.1.3 Time interval analyzers

One of the goals in measuring medium noise is to find a way to measure and compare the noise of various media on a certifier rather than in the completed disk drive. However, for practical purposes one needs measurement techniques which predict error rate performance in the completed disk drive. In addition, one would like measurements which also aid in diagnosing error rate problems. A full error rate test of a drive takes a considerable length of time and, if a head/disk combination in that drive fails, simple error rate provides no help in diagnosing the cause. A time interval analyzer can be used to distinguish between the effects of transition (medium) noise and electronics (readback) noise in a way that correlates well with phase margin or error rate results in a finished drive.

Because the magnetization noise in thin film media resides primarily in the transitions (and is often called "transition noise") it manifests itself as an uncertainty in the detected location of each transition. This uncertainty is just the standard deviation of the detected transition positions about their mean position. This mean position is defined as the average time between transitions, and can be compared to the expected time based on the frequency at which they were written. As mentioned at the beginning of the chapter, mathematically the *noise* in a recording channel can be represented by the *variance* in the detected location of the recorded transitions (in time or space). If, as is usually the case, there are basically two contributors to noise (transition or write noise and electronics or read noise) then the total noise is the result of these individual contributing noise sources as the equation below shows:

$$\sigma^2_{total} = \sigma^2_{write} + \sigma^2_{read} \tag{8}$$

where the subscripts refer to the several noise sources. This standard deviation, σ, is called the "jitter" of the transition, or the "transition jitter noise." It can be expressed in nanoseconds but is better expressed in nanometers (nm), since it is the uncertainty of the transition's location in space along the recorded track. Also, expressing this noise in nanometers is independent of the head/disk relative velocity and is thus more easily compared to other head/media combinations. It must be remembered, however, that the transition jitter is *not* a simple function of the total noise power in a given disk's transitions. The jitter is also determined by the sharpness of the readback pulse induced by transitions on that disk and by the input filter's shape and bandwidth. Thus, two media which have the same measured noise power with a given head can have quite different jitter noise if their pulse widths or input filters differ. Also, the same disk measured with two different heads can have two different results for jitter noise if, for instance, the heads have different recording resolution. So, all comparative measurements must be evaluated with great care to try to ensure valid comparisons.

3.1.4 Phase margin analyzers

Progress in measurement usefulness is made by the use of a phase margin analyzer. Phase margin analyzers write a user-specified pattern of data around the track and then read that data back with a modified disk drive detection channel. During readback, a phase margin analyzer changes the width of the timing window used to determine if there is a 0 or a 1 in each clock cycle and, by comparing the recovered data with the originally written data, measures the bit error rate as a function of the timing window width. The curve thus obtained shows an increase in the number of errors as the window width is reduced from the nominal full width for the channel. Even though only a few tens of millions of bits are usually tested (which takes just a few seconds to a minute), the curve can be extrapolated by well-developed techniques to compute how much of the full timing window will be available at the desired high levels of data reliability, commonly 1 error in 10^{10} bits, or a bit error rate of 10^{-10}. Used to test surfaces in a drive, it is much faster than full error rate testing and, used judiciously, can provide some information to distinguish causes of margin failure. It can also be used to test individual heads and disks on a spin stand or certifier. Phase margin testing is a rapid and effect method for analyzing the impact of different recorded patterns on probability of error.

Moreover, the predictive value of phase margin tests on a spin stand is high. However, it is difficult to unambiguously distinguish margin failures due to medium noise from those due to electronics noise. All *noise* sources tend to act in concert to make the margin worse. Therefore, despite the great usefulness of phase margin analysis, if a head/disk combination exhibits inadequate margin additional non-phase margin testing will be necessary to determine the source of the problem.

3.2 Measurement Dimensions

The dimensional quantities used to report medium noise vary somewhat from experimenter to experimenter and paper to paper depending on the kind of instrument used to measure noise, the purpose of the experiment or the application of the noise measurements, and the tastes of the experimenter. For instance, one experimenter will report medium noise as volts/$\sqrt{Hz}$, another as (volt)^2, and still another as nanometers, to mention a few. This confusion of measurement quantities or dimensions employed for medium noise is an inevitable byproduct of the complexity of the subject and of the experimenters' goals. The dimensions may be categorized in three general ways: (1) frequency domain dimensions, e.g., (volt)^2 measured with a spectrum analyzer; (2) time domain dimensions, e.g., nanometers of jitter in transition location; and (3) magnetic dimensions, e.g., (amperes)^2 for the intrinsic NPSD of magnetization fluctuations (see Eq. 6). These dimensions are all interrelated because they represent different ways to measure the same phenomenon. Frequency domain dimensions have been most common because, until recently, most media noise was measured using a spectrum analyzer.

In the last several years people working in the field of medium noise research have been striving to develop a common way to represent the medium noise of various media in a way which makes the practical impact of a given noise level immediately evident. The problem with simply quoting the total noise power of different media is that this number alone does not correlate well with error rate or phase margin measurements in the time domain, which are measurements which predict error rate performance in a particular disk drive with a particular kind of head and channel. In fact, it is not hard to find examples of two disks, one of whose normalized total noise power is lower than the other's, yet the one with the higher absolute noise exhibits better margin performance in a drive. Also, signal-to-noise ratio, no matter how it is defined, has little or no predictive value for thin film media. The reason is found in the relationship between noise power (or voltage) and standard deviation of the actual location of recorded transitions about their mean. This relation was first derived by Tahara, et. al. (Tahara, et. al., 1976) and reduced to a simpler form by Katz & Campbell (Katz & Campbell, 1979), who wrote that the standard deviation of transition locations in time about their mean time of detection (for a peak detector channel) was related to the noise and to the shape of the isolated readback pulse by the equation:

$$\sigma = \frac{\left(\dfrac{dn}{dt}\right)}{\left\{\dfrac{d^2 V_O(t)}{dt^2}\right\}_{t=0}} \tag{9}$$

where [dn/dt] is the rms noise voltage as measured at the output of the differentiator (before the zero-crossing detector) and the denominator is the curvature of the isolated pulse at the peak. Both noise and pulse curvature will, of course, also be affected by the shape and bandwidth of the input filter or equalizer. "σ" represents the uncertainty in transition location and is usually called the "time jitter" or "transition jitter." It is related by the error function to the probability of error in detecting a single bit , that is, the probability of error as a function of location t' in the decoding window of the channel is:

$$prob.\,of\,error = p.e.(t') = k \int_{t=t'}^{\infty} e^{\frac{-(t-offsets)^2}{2\sigma^2}}\,dt \tag{10}$$

3.3 Normalizing the measurements

Often the experimenter must compare the noise of different types of media. When the media being compared do not have the same amount of magnetic material, or $M_r\delta$'s, they will not produce the same signal and noise levels. Therefore, in order to adequately compare the noise performance of the different media their noise and signal levels must be normalized to some common basis. It is natural to normalize the total noise power, P_n, from Eq. 7 by $(M_r\delta)^2$ in order to compare different media measured with the same head. If different heads are used to measure different media the normalization is more complex, and in principle would require dividing P_n by $(\mu_0 N \varepsilon v M_r\delta)^2 w$ (see Eq. 5). In practice it is difficult to accurately measure the head efficiency, ε; The most correct way to do the normalization is to divide the noise voltage by $M_r\delta$ or the noise power by $(M_r\delta)^2$, where δ is the thickness of the recording film. This is because the reciprocity equation, which governs the relationship between the magnetic properties of head and medium and the magnitude of the output signal, depends on $M_r\delta$. Sometimes normalization by the peak-to-peak amplitude of the isolated pulse signal, V_0 has been used in the literature. This is satisfactory as long as the experimenter is comparing media whose transition parameters are the same, so that the relative signal amplitudes of different disks are really determined by their $M_r\delta$'s. However, it is technically incorrect and will frequently lead to incorrect results because V_0 is affected by the sharpness of the recorded transition as well as by the amount of magnetic material present ($M_r\delta$). Of course, V_0 is easier and more convenient to measure than $M_r\delta$, and has the distinct advantage of being a non-destructive measurement; cutting up your experimental media for a vibrating sample magnetometer (VSM) measurement does make it difficult to use them again! On the other hand, it can be noted that it is not necessary to cut up the media in order to gain information about it remanence. A paper by Silva and Bertram (Silva and Bertram, 1990) , describes a non-destructive *in situ* technique for measuring medium magnetization-thickness products via a recording measurement. This technique is useful, but the method they describe requires some modification. This is because Silva and Bertram overlooked the effect on this remanence measurement of the varying demagnetizing fields produced by the recording process with varying write current. J.-G. Zhu discusses this later in this volume in his chapter on micromagnetic modeling. The error can be substantial, so great care must be exercised in using non-destructive recording methods to determine MH loop properties of media. The most certain way to obtain $M_r\delta$ for normalizing noise measurements is still to cut up the disk and measure it with a VSM.

4. General Characteristics

In particulate media the noise is pretty much uniformly distributed along the track, not exhibiting any "preference" for the transition over the dc-saturated regions between transitions. The character of medium noise in particulate media can be quantitatively accounted for by the combined effects of the statistics of independent particles, fluctuations in the local distribution of the particles and clumping of the particles into larger aggregations (Thurlings, 1980). The situation is more complex for noise in metal film media.

It is difficult to quantify the medium noise from oscilloscope traces because the noise is usually greatest at fairly low frequencies compared to the signal waveform (see Fig. 1). Therefore it is usually characterized and studied through the use of a spectrum analyzer. As mentioned in the historical introduction to this chapter, the total noise power measured in thin metal film media is a function of the recording density; as shown in Figure 2 medium noise increases as the recorded density increases. Baugh, et al, (Baugh, et. al., 1983) first understood the reason for this increase, based on an analysis of what spectrum analyzers actually measure. A spectrum analyzer measurement of voltage or power at a given center frequency takes a certain amount of time – as much as several milliseconds, depending on the instrument's settings. The resulting power measurement is the average signal or

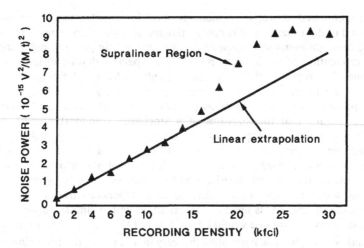

Figure 2: Typical curve of total noise power versus recording density for CoPt media, showing linear and supralinear regions.

noise power within the resolution bandwidth during the measurement interval. Because a recording disk is moving rapidly past the head, during each measurement thousands of transitions pass the head.[2] Each data point in a noise spectrum is therefore the average of noise at that frequency over a section of track containing thousands of transitions *and the magnetically saturated regions between transitions.* Since the noise power increases with recording density it is not statistically "stationary." That is, the noise is not the same from place to place along the track (Papoulis, 1965, pp. 300ff). The observed linear increase of noise power with increasing recording density corresponds with the increasing fraction of the track that is occupied with transitions. Therefore, medium noise is greater in the transition regions than in the saturated zones between transitions. Later experiments by E. Yarmchuk (Yarmchuk, 1986) in which the signal from many isolated transitions was captured by single-shot digitizing oscilloscope measurements and averaged using special techniques to obtain both down-track and cross-track signal shapes corroborated this model; the voltage noise in the transitions was always much greater than that between transitions.

---

[2] Typical spectrum analyzer settings useful for measuring medium noise in high performance disks might be: frequency span=20 MHz, resolution bandwidth=30 Kz, video bandwidth (post-detection filter)=30 Hz. With these settings the sweep time of a typical spectrum analyzer would be 70 seconds for 1001 data points. Each data point takes a measurement time of somewhat less than 70 milliseconds, but more than 50 msec. Increasing the VBW to 300 Hz decreases the measurement time by a factor of 10, to approximately 5 msec. At typical head/disk speeds on 95mm media of 5-10 m/sec each measurement covers a distance of *at least* 2.5 cm and as much as 50 cm ! Since transitions are typically spaced at most a few micrometers apart, it is easy to see the extreme averaging that occurs.

As mentioned above, since medium noise in recorded metal film media is concentrated in the transitions it is not "stationary" in a statistical sense. This means that noise measurements with a spectrum analyzer are, in general, not statistically valid for calculation of the intrinsic noise power spectral density of magnetization fluctuations. The reason is that a necessary condition for the definition of a power spectrum as the Fourier transform of the correlation integral (Eq. 2-6) is that the statistical process involved be *stationary* (Papoulis, 1965, Ch. 9, Ch. 10 Section 10.3). "Stationary" means that the statistics of a stochastic (random) process are not affected by a shift in time. Applied to a spatially stochastic process such as medium noise, stationary means that the noise statistics do not depend on where the measurement is made along the track. A track recorded with separated transitions has noise which does vary with location and the statistics of the noisy transitions are different from the statistics of the saturated regions between transitions; thus, the noise is not stationary. (This must not be confused with a simple duty-cycle effect, for which the noise magnitude would be too low but the statistical form of $S_\phi(f_x)$ in Eq. 5 would be correct. The noise statistics vary in and out of the transition regions, making a track recorded with isolated transitions truly non-stationary.) The term $S_\phi(f_x)$ in Eq. 5 is not a statistically valid quantity, since it mixes the noise from statistically different regions of the track. In that sense, Eq. 5 *is not valid* for transition noise if the transitions are isolated. However, fortunately there is a way around this problem. That is to record the track with a pattern at sufficiently high density to effectively fill the track with transitions. Then the magnetization noise along the track *is* stationary, Eq. 5 holds and analysis of the intrinsic magnetization noise using Eq. 5 becomes possible. Experimentally, this means to record the track at the recording density which produces the most total noise power, at the peak of the curve in Fig. 2.

Before introduction of thin metal film media the common figure of merit used to predict the impact of medium noise on a recording channel was the signal-to-noise ratio (SNR). This was conventionally defined as 20 times the logarithm of the zero-to-peak signal voltage at some recorded density divided by the broadband rms noise voltage measured for a dc-saturated (erased) track. SNR depended on the recording density only to the extent that the signal did. From the SNR it was possible to predict the contribution of medium noise to channel error rate and to compare different media. Since medium noise in thin film media is concentrated at the transitions, any definition of SNR that uses dc-saturated noise (*much* lower than the noise in the transitions, as Fig. 2 shows) will yield a hopelessly optimistic estimate of the medium contribution to error rate. For peak detection channels, the best approach seems to be to use Eq. 9 and 10 to calculate the jitter of transitions (or, better yet, to measure the jitter directly) and the resulting error rate contribution; for sampled voltage detection there is, as yet, no universally recognized way to represent medium noise in a useful way. If Eq. 9 is used to compute transition jitter, the noise voltage used must be the maximum noise; moreover, it should be referred to the output of the differentiator for peak detection channels (i.e., the noise spectrum at the head terminals should be differentiated).

The realization that noise is concentrated at the transition locations explains the linear portion of the curve in Figure 2. It assumes that the average noise power per transition is a constant, independent of recording density. However, in nearly all types of thin film media it is observed that the noise power increases *faster than linearly* as the density continues to increase, peaking at some higher density. Details of the explanation for this effect will be found in the next section; for now, suffice it to say that a comprehensive understanding of medium noise requires an understanding of this effect.

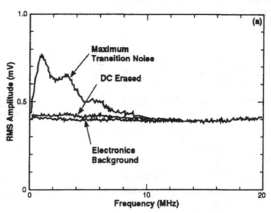

Figure 3a: Measured noise spectra for maximum transition noise and dc-saturated noise on one disk.

Consider now a closer look into the character of this transition noise. Figure 3a shows how the observed spectra look on a spectrum analyzer for maximum transition noise and for dc-erased noise[3] on a single disk with a thin film head and Figure 3b shows the observed spectra for a low noise and a high noise disk as measured with a ferrite head (the results are the same for a thin film head, with some differences in detailed spectral shapes attributable

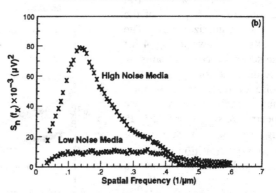

Figure 3b: Measured noise spectra for maximum transition noise from a high noise and a low noise disk.

---

[3] The terms "dc-erased" and "dc-saturated" are used interchangeably to mean a track which has been prepared by applying, for one complete revolution of the disk, a constant, unidirectional head field sufficiently large to saturate the medium and erase any pre-existing transitions.

to the finite pole tip lengths). The first thing to note is the expected differences in magnitude of the various curves. Also, however, note that for each pair, and indeed, for all four spectra, the overall shapes of the curves are similar. That is, as frequency increases noise power increases rapidly to a peak then falls slowly back to the background. The extra peaks in Fig. 3a are due to the effects of the finite pole tip lengths of a thin film head compared to a ferrite head, as in Fig. 3b.

From these measurements it would be quite difficult to determine if there are any significant differences in the noise statistics of the four recorded states. Recall, however, our discussion from Sections 3.1 and 3.2 about the use of a spectrum analyzer to measure noise. Using Eq. 5, spectra such as the above can be analyzed to determine the intrinsic noise power spectral density curves, $S_\phi(f_x)$, of the media magnetization. The results are shown in Fig. 4 for the several recorded states on one disk and for the transition noise of high and low noise media whose raw noise spectra are shown in Fig. 3. In Fig. 4a, S_ϕ is shown for maximum transition noise and dc-erased noise from a Cr/CoPt disk; in Fig. 4b, we show S_ϕ for the maximum transition noise of two Cr/CoP disks, one with relatively high noise and one with relatively low noise.

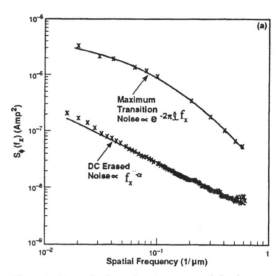

Figure 4a: Magnetization noise power spectral density curves for maximum transition noise and dc-erased noise on one disk.

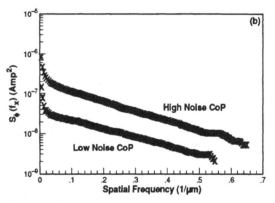

Figure 4b: Magnetization noise power spectral density curves for maximum transition noise from a high noise and a low noise Cr/CoP disk, respectively.

In Fig. 4a it is immediately clear that there are differences in the functional dependence of the spectral densities which were not apparent before removing the filtering effect of the recording head. In fact, all three transition noise spectra follow very close to a $S_\phi(f_x) \propto e^{-\alpha f_x}$ dependence, while the dc-erased noise follows a $S_\phi(f_x) \propto f_x^{-\alpha}$ dependence. Work by Baugh, et. al. (Baugh, et. al., 1983) reported similar results in 1983. Many workers since have shown that recorded transitions tend to exhibit a complex sawtooth pattern across the track, with the sawtooth amplitude and angles characteristic of the type of media and, probably, the presence and strength of quantum mechanical exchange coupling between grains in the film plane.

If one can draw extensions from one field to another, Majumdar and Bhushan have shown for surface roughness of recording media and other "smooth" surfaces that an exponential Fourier power spectrum of the surface of the form $f_x^{-\alpha}$ is caused by fractal geometry of the surface itself (Majumdar and Bhushan, 1990). That is, the roughness is scale-independent, appearing the same at many measurement length scales. The point here is not that medium noise is necessarily caused by roughness, but that a power spectral density of some physical quantity that follows a power law like $f^{-\alpha}$ suggests a possible fractal character. It may be that the similar power law dependence for dc-saturated noise is evidence that the dc-saturated noise is a fractal quantity. The question naturally arises as to whether transitions have a fractal character. Che and Suhl (Che and Suhl, 1990) have modeled transition formation and spectral dependence using the principles of self-organized criticality, one of the mathematical generators of fractal character in many physical processes. A fractal character of the transition would not really be surprising, since the transition is formed by some stochastic (random) process during magnetization reversal, and geometric objects of fractal character are also often formed by stochastic processes. It may provide some additional insight into how transitions form and how to design media with lower noise (magnetization fluctuations). However, the few spectral densities of transition noise reported to date follow exponential functions of spatial frequency, not power laws, and are therefore evidently not fractals. It is an interesting area of research which may provide further insight into transition formation and other types of medium noise.

One additional feature of the noise power spectral density (NPSD) of transition noise should be noted. That is that the NPSD is the Fourier transform of the autocorrelation function of a statistical process. The autocorrelation function is a measure of the range over which the statistical fluctuations are not independent of one another, i.e., over which they are correlated. Interestingly, an exponential NPSD when inverse-Fourier transformed gives rise to an autocorrelation which is Lorentzian in shape with a characteristic length (a "correlation length") associated with it. Mathematically,

$$S_\phi(f_x) = F\{R(x)\} \tag{11}$$

$$\text{and} \quad R(x) = F^{-1}\left\{S_\phi(f_x) \propto e^{-2\pi \ell_x f_x}\right\} \quad \propto \quad \frac{1}{\left[1+\left(\frac{x}{\ell_x}\right)^2\right]} \tag{12}$$

where R(x) is the autocorrelation of the magnetization noise; $F\{$ $\}$ and $F^{-1}\{$ $\}$ are the Fourier and inverse Fourier transform operations; and ℓ_x is a correlation length. From Eq. 11-12 we may infer that the exponential dependence of $S\phi$ arises from noise or fluctuations which are extended in space along the track, with a characteristic length. Since the transition noise "fills" the transition region, which is itself characterized by a transition width parameter, a, it is tempting to see if the noise correlation length is related to the Williams-Comstock transition parameter. This would be consistent with a noisy transition whose noise intensity follows along with the magnetization reversal - like a sawtooth transition form would do. Baugh, et. al. (Baugh, et. al., 1983) studied this effect and found that the correlation length calculated from the measured transition NPSD correlates extremely well with the transition parameters, a, calculated from the Williams-Comstock model, for a number of media alloys and a wide range of H_c and $M_r\delta$. Figure 5 here shows the results of Baugh, et. al.'s investigations. It is clear that there is indeed a good correlation between ℓ_x and a, lending support to a connection between the two. Belk, George and Mowry (Belk, et. al., 1985) reported on similar investigations of this effect, although they looked for correlation of "$2\pi a$" with the inverse of the density where transition noise peaked, which should be related to the total width of transitions (see the next section and Fig. 6). For both research teams this correlation held for a wide variety of thin metal film media, both sputtered and

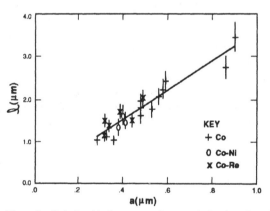

Figure 5: Relationship between noise correlation length and calculated transition parameter. (Baugh et. al., 1983)

plated and both binary and ternary cobalt alloys. Little attention has been given in the literature (but is in Chapter 4, Section 2.2.2) to these aspects of medium noise.

5. Supralinear Noise Region

Most measurements of transition noise power as a function of recording density result in a curve like Figure 2. As discussed in the previous section, for low to moderate recording density the noise power increases linearly, but as the density continues to increase the noise power typically grows faster than linearly, finally peaking and declining slowly for very high recording densities. The understanding that noise measured by a spectrum analyzer or rms-voltmeter is an average of the noise over some portion of the track circumference, coupled with the understanding that most medium noise is localized in the transitions, leads to a prediction of linearly increasing noise power with density until transitions begin to overlap. This simple model cannot account for the commonly observed supralinear region of Figure 2. The result of this region is that transition noise is worse for closely spaced transitions than for isolated ones. It is therefore important to understand what causes this phenomenon.

Although first noted for a variety of thin film alloy media by Baugh, et. al. in 1983 (Baugh, et. al., 1983), the earliest systematic investigation of this effect was by Belk, et. al. in 1985 (Belk, et. al., 1985). They examined a collection of media including four cobalt alloy thin film media, one particulate disk and one CoCr perpendicular recording disk. They found that the total noise power increases faster than linearly for all the thin film longitudinal media but not for the particulate disk or

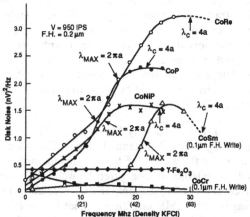

Figure 6: Disk noise vs. write frequency (or density) for various media. The parameters λ_{max} and λ_c are calculated using the Williams-Comstock formulas to evaluate the transition parameter, a. (Belk, et. al., 1985)

the perpendicular disk. (See Figure 6) The media chosen exhibit this effect in particularly dramatic fashion, especially the CoSm media. The parameters λ on their graph mark the wavelengths predicted by Belk's model for the location of the peak in noise power and essentially show that the positions of the noise maxima correlate well with the transition parameters; this was to be expected if the model proposed by Baugh et. al. (1983) was essentially correct, namely that the noise maximum occurs when the track is filled with non-overlapping transitions. It is significant that in both these papers the degree of supralinear effect varies substantially depending on the alloy and deposition conditions, suggesting that the cause lies in some micromagnetic property of the media.

Noting that the supralinear noise increase suggests that the individual transitions become somehow "noisier" as their density (and hence proximity) increases, Belk, et. al. and others have attempted to explain the effect in terms of magnetostatic interactions between adjacent transitions as they are written closer and closer together. Belk, et. al. (1985) added to the model of Baugh, et. al. (1983), in which the observed transition noise is caused by a jitter in the positions of written transitions. Belk explained the anomalous noise increase in the supralinear region by a model in which individual transitions (assumed to be sawtooth across the track) do not become more jagged but in which the transition walls repel one another, so jitter in the position of one transition pushes the next transition in the same direction, in addition to its own jitter. In their model this type of correlation between the jitter on adjacent transitions leads to additional media noise for closely spaced transitions (Belk, et. al., 1985). They concluded that the transition correlations must be negative (repulsive) because of the observed supralinear noise. Note, however, that this is a counter-intuitive result. Each transition can be considered as a jagged line of magnetic poles where adjacent transitions consist of poles of opposite polarity. It is therefore likely that any correlation in jitter between adjacent transitions would be attractive (positive correlation) rather than repulsive (negative correlation).

Subsequent work by Madrid and Wood (Madrid and Wood, 1986) and by Tang (Tang, 1986) extended and refined the ideas of correlations between transitions as the source of the anomalous noise. Madrid and Wood found that, indeed, correlation in the jitter between adjacent transitions could explain the observed extra noise, but only if the correlation resulted from adjacent transitions jittering in the SAME directions. That is, the interaction between transition walls would have to be repulsive as Belk, et. al. had found, for noise to increase beyond the linear dependence on transition spacing. They also found that the total noise would DECREASE if transition jitter correlated in opposite directions (attracted one another). However, in contrast to Belk, et. al., Madrid and Wood found considerable evidence that despite the existence of supralinear noise, the actual correlation between transitions was attractive, which would result in less noise rather than more. They found this evidence in careful analysis of experimental noise spectral densities. As a result, they suggested that the observed supralinear noise must therefore be due to an increase in the intrinsic noise of individual transitions; it could not be explained by correlations. Madrid and Wood also suggested that demagnetizing fields between transitions would result in attractive correlation of transition jitter -- essentially that opposite magnetic poles attract through their demagnetizing fields. Tang, on the other hand, used a very general definition of autocorrelation function for transitions to include the possibility of width fluctuations as well as position fluctuations. Analysis of experimental data from a single disk using this method led Tang to the conclusion that medium noise is dominated by position jitter for widely spaced transitions, but for increasingly closely spaced transitions there is also a contribution from transition width fluctuations, essentially a producer of increased noise per individual transition. Indeed, in Tang's results the width fluctuations dominate the noise power for transitions closer than a transition width apart. This provided additional evidence for increase in intrinsic transition noise at close spacings.

A paper appearing in the same journal issue by Arnoldussen and Tong (Arnoldussen and Tong, 1986) provided pictorial evidence which supported the results of Madrid and Wood and of Tang. They noted that transitions of opposite polarity should attract one another, producing just the opposite jitter correlation from that proposed by Belk, et. al. In support of their proposal, Arnoldussen and Tong showed Lorentz micrographs of transitions in Fe-Co-Cr media at various transition spacings. Although the media used was not high performance media, the consistency of the noise power results for all kinds of thin film longitudinal media leaves little room to doubt the validity of their result. Specifically, they observed that as sawtooth transitions approach one another adjacent portions of the sawtooth attract one another. This results in both an attractive correlation in the net position jitter and, because the interaction results in local elongation of the zigzags, the transition fluctuations are amplified as

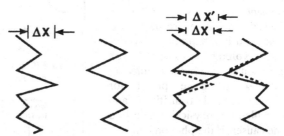

Figure 7: Sawtooth elongation and bridging as transitions approach each other. Δx represents variance, or jitter. (Arnoldussen & Tong, 1986)

they approach one another. Figure 7 is a diagram from their paper showing what the Lorentz micrographs revealed about the interactions between approaching transitions. Thus the intrinsic noisiness of closely spaced transitions is greater than that of isolated transitions. In comparing their results with those of Belk, et. al., Madrid and Wood and Tang we note that Arnoldussen and Tong showed that correlation between closely spaced transitions was indeed attractive (i.e., they jittered in toward each other, in opposite directions) and that the intrinsic noise of the transitions increased at close spacing. Since Madrid and Wood had shown that attractive correlation would REDUCE medium noise, the observed supralinear noise could not arise from jitter correlations: the excess transition noise had to be caused by the increase in intrinsic noisiness of closely spaced transitions.

At that point, it remained for someone to put the observed effects and the ideas about transition correlations on a solid theoretical footing. The first approach to this was by Alex Barany, a graduate student under Prof. Neal Bertram at the University of California, San Diego, in his Ph.D. thesis and associated paper (Barany and Bertram, 1987). Without assuming anything about the details of transition structure (i.e., whether sawtooth or not), Barany modeled the noise process in longitudinal media assuming each transition's position fluctuation is affected by the previous transition's position fluctuation through the demagnetizing field. The dominant effects he found are an increase in the inherent noise (jitter) in each transition's position and a broadening of the transitions as the recording density increases. Although flawed in some respects (as discussed in Chapters 4 and 5 of this volume), Barany and Bertram's model did predict an increase in noise power for closely spaced transitions and provided key physical insight into the causes of the phenomena.

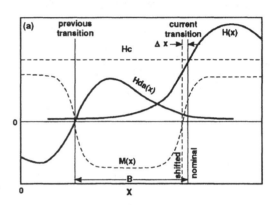

Figure 8a: Effect of adjacent transition on the write process. (Barany & Bertram, 1987)

The basic effects are illustrated in Figure 8. Fig. 8a shows the effect of the demagnetizing field of an existing transition on the position of the next on to be written. That effect is an attractive interaction which moves the new transition closer to the previous one than its location would have been in the absence of the first transition. In

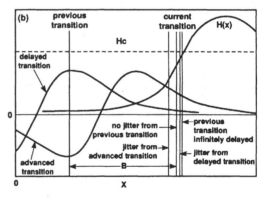

Figure 8b: Effect of adjacent transition jitter on write process.

the presence of jitter in the position of the first transition, as in Fig. 8b where transition jitter is represented by "advanced" or "delayed" transitions, the second transition's location is affected to a different degree depending on the "jittered" position of the first. Thus, *additional uncertainty* is introduced in the location of each new transition because of the fluctuations in the locations of preceding transitions. This additional uncertainty is *added* to the initial intrinsic position uncertainty of each new transition, presumably caused by microscopic variations in media properties and microstructure. Thus, higher recording density increases the intrinsic position jitter of transitions through their interactions with demagnetizing fields.

The most comprehensive analysis of the causes of medium noise in general and noise of closely spaced interacting transitions in particular have been done by Zhu and Bertram (Zhu and Bertram, 1988a and 1988b) and Zhu (Zhu, 1991). Since an entire chapter (Chapter 6) in this volume is devoted to this subject, little will be said here. Suffice it to say that the supralinear noise effect is now quantitatively understood through numerical micromagnetic modeling, confirmed by experiment. Basically, the supralinear noise is the result of the combined effects of the demagnetizing fields between adjacent transitions and the quantum mechanical exchange interactions between grains in the recording film. This leads to larger fluctuations and increased short-range interaction between transitions than is the case with non-exchange coupled films, and thus increases the noisiness of closely spaced transitions (See Chapter 6, Section 4.2).

6. Types of Medium Noise

6.1   Recorded Magnetization States

The focus up to now has been on the measurement and characteristics of the noise associated with written transitions in longitudinal thin film media. Magnetic materials are, however, quite complex and a given material may exist in any number of magnetic states. Associated with each magnetic state of the recording medium is noise, or fluctuation of the magnetization in that state. Of course, only a few of the possible magnetic states are of any practical importance for recording. However, those few encompass more than just the noise within a recorded transition. The magnetization noise of several other magnetic states is of importance, either because it may affect the performance of the head/medium system or because it can be used as a tool to gain insight into the causes of medium noise or some other aspect of head/medium performance, such as head efficiency.

Figure 9 is a typical hysteresis loop for thin film media, showing several of the magnetization states of interest in the performance or study of recording media.

88

(a) This magnetization state corresponds to the center of a transition <u>while the transition is being formed</u>, where the recorded track to either side is in state (1) and its equivalent saturated state on the other side of the loop. The transition center then relaxes to state (b) upon removal of the writing field of the head. It is very similar to a state (b) as reached from state (3), except that

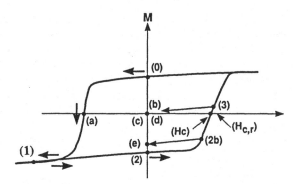

Figure 9: Typical hysteresis loop for thin film media. Marked points correspond to magnetization states mentioned in the text.

in the dc-demagnetized state the entire length of the track is in the same state, while the transition center is in a demagnetized state but is surrounded by regions in states (0) and (2).

(0), (2)   This is the <u>dc-saturated or dc-erased</u> state, found between transitions in the center of a bit cell or in a stretch of track uniformly saturated (erased) by a large field.

Three other demagnetized states exist, two reached by paths difficult to show in the Figure but distinct from one another.

(b) First is the <u>dc-demagnetized</u> state. It is reached by following the path (2) → (3) → (b) and leaves the medium in a state having zero magnetization, but with a different environment from the zero magnetization state at the center of a transition.

(c) Second is the <u>as-deposited or virgin</u> state, formed as the film grows during its deposition. This will not be quite demagnetized if (as is generally the case) the film growth occurs in a magnetic field, but unless the field is larger than 100-200 Oersteds the virgin state will be very close to zero remanent magnetization (demagnetized)

(d) Third is the <u>ac-demagnetized or ac-erased</u> state. This is reached by cycling the applied field from positive to negative and back, beginning well above the coercivity, H_c, and slowly reducing the field to zero while it is cycling. This state can be approximated by writing at very high frequency, but this is only an approximation of the true ac-demagnetized state.

In the following paragraphs the key characteristics of each of the magnetization states (except the transition center, which has already been discussed as transition noise) will be described. Note that the *write current* through the recording head produces the *magnetic field* from the head in the plane of the medium. The current and field are thus essentially interchangeable terms in the following discussions.

6.2 On-track dc-demagnetized noise (reverse-erased noise)

Most of the attention regarding medium noise has been focused on transition noise, and rightly so. However, there are certain difficulties inherent in measuring transition noise, such as the need to remove from the measurement the signal and its harmonics. This requirement complicates the measurement and also makes the interpretation of the noise spectrum more difficult, because the process of removing the signal is more of an art than a science. By referring back to Figure 1 the reader can see the presence of the signal peaks and harmonics on the raw spectrum analyzer traces. Clearly, to measure medium noise these signal peaks must be removed from the data. The simplest way to do this is to have the computer locate these peaks and delete the data points in a region on either side of the peak, to ensure removing all the signal energy down to the noise baseline. Then the computer can make a linear interpolation across the gap to fill in approximate values of the noise data points under the peaks. Since the spectrum usually changes rather slowly under the peaks this approximation will be quite close. There are some pitfalls to avoid, however. First, a sufficiently wide stretch around each peak must be deleted in order to be sure of excluding all the signal power from the result. Second, deleting too wide a region about the peaks can lead to a situation where most of the final interpolated spectrum is actually the linear interpolation sections rather than the true noise data; this will distort the shape of the spectrum and, in extreme cases, the value of the integrated noise power as well. Note that the most extreme case would be one in which the interpolation regions overlap and no actual noise data is left! This can actually occur if the recording signal's frequency is low compared to the width of the region that will be deleted around the peaks. Care must be taken to properly remove the signal power, especially at low recording density. Of course, this will not be a problem at the fairly high recording density needed to measure the transition noise (recall that this noise is the *maximum* noise as a function of recording density).

Moreover, in principle one would like to be able to measure certain properties of the recording medium and the head without cutting up the disk or risking destroying the head. A recording measurement which would extract the coercivity, for instance, or the head efficiency would be quite useful. The method of dc demagnetization noise provides a useful, non-destructive measurement tool for estimating transition noise, characterizing media coercivity with spatial resolution and measuring certain properties of the recording head. Properly used, it also provides a sensitive probe for determining the statistical mechanism producing dc demagnetization noise, from which inferences may be drawn about aspects of the mechanism of transition noise. The technique is therefore worth some detailed attention.

In 1986 this new measurement technique was introduced by two sets of investigators working independently (Bertram, et. al., 1986 and Aoi, et. al., 1986). Both groups showed that the medium noise power varied in a characteristic way if the track was prepared as in case (b) above. That is, a recording track is first saturated by applying a constant current to the head in a fixed direction, resulting in a constant magnetic field at the medium. With the head field off the medium relaxes to the remanent state (2) shown in Figure 9. Then a smaller steady current in the opposite direction is used to excite the head, its field moving the medium magnetization partway along the hysteresis curve from (2) to (3) in Figure 9, such as to point (2b). When the head is de-energized its field falls to zero and the medium relaxes to the less-saturated remanent state (e), at which point the noise power is measured. This process of preparing the track in state (2) and applying a <u>reverse</u> dc magnetic field is repeated, driving the medium gradually along the entire right branch of the hysteresis curve. The measured noise power traces out a curve like Fig. 10, showing a distinct peak at some reverse current.

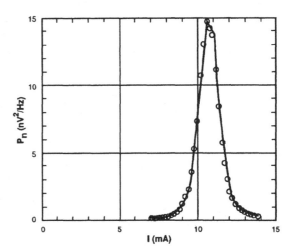

Figure 10: Noise power versus reverse dc excitation current. (Tarnopolsky, et. al., 1989)

Both Aoi, et. al. (1986) and Bertram, et. al. (1986) showed that the value of the current at the peak in noise power is near the current which produces a field at the medium surface equal to the "remanent coercivity." The remanent coercivity, $H_{c,r}$, is slightly greater than the normal coercivity, H_c, and is that applied field required to produce a magnetic remanence of zero after the field is removed (point (3) on Fig. 9). Aoi argued that case based on a linear correlation between the current producing the noise maximum, I_{pk}, and the normal coercivity for a number of media, while Bertram argued for $I_{pk} \Rightarrow H_{c,r}$ based on an observed near equality between I_{pk} and two other experiments. Although the case was plausible, in neither case was it really proved that I_{pk} produced a field of $H_{c,r}$ at the medium; in fact it does not exactly do so. Two years later Tarnopolsky, et. al. (1989) proved the relationship between I_{pk} and $H_{c,r}$ quantitatively. They showed definitively that I_{pk} produces a field at the medium that is nearly but not exactly $H_{c,r}$. The difference between the field I_{pk} produces and $H_{c,r}$ is due to demagnetizing fields set up in the medium by an effective magnetic charge pinned under the leading pole of the recording head as it is switching the medium from a near-saturated remanent state (like (0) and (2) in Fig. 9) to the state which will be exactly demagnetized once the field has been removed. The error depends on the demagnetizing factor $(4\pi M_r \delta / H_{c,r})$ of the medium and can be as high as 20%. Tarnopolsky, et. al. calculated the demagnetizing field and its effect using a self-consistent model of the recording process. The calculation is described in their paper.

Several other results have come out of the work on dc demagnetized noise. First, there appears to be a reasonably close agreement between the magnitude of the dc demagnetized noise and the transition noise for many kinds of media. Aoi, et. al. (1986) examined that question for eight different kinds of media and found the two types of noise to be equal within 10-20%; a similar correspondence has been noted in passing by other workers since then (e.g., Murdock, et. al., 1990; Tarnopolsky, et. al., 1991) . Aoi goes so far as to suggest that dc demagnetization noise can be measured *in lieu* of transition noise, since it is easier and quicker to measure. This seems unwise despite the close agreement in magnitude. There is no *a priori* reason to expect that dc demagnetized noise and transition noise would be equal, although they might be expected to be close since, as Fig. 9 shows, the magnetic states of the medium which give rise to the two types of noise are similar but not identical. For the noise powers to be nearly equal suggests that the size, extent and distribution of the magnetization fluctuations are very similar in the two states. Zhu discusses this issue in Chapter 6, Section 6 of this volume. Perhaps there is a subject here for future investigation!

In addition to the similarity of noise magnitudes, investigators of the dc demagnetized noise phenomenon have shown that it is extremely useful for other quantitative measurements on heads and media. In their 1986 paper, Bertram, et. al. (Bertram, et. al., 1986) showed that dc demagnetization noise can be used to calculate head efficiency from I_{pk} or, in some cases with textured media, to calculate an estimate of the texture roughness, based on modulation of the noise spectral density by flying height variations. In addition, slow spatial variations in coercivity around a track can be qualitatively observed by properly triggered zero-span spectrum analyzer measurements, providing a tool for monitoring the uniformity of the film deposition process over the disk.

Finally, insight into the physical mechanism of dc demagnetization noise has been gained by careful measurement of the dc demagnetization noise vs. reverse excitation current curve, coupled with appropriate modeling. In three papers it has been shown that the shape of this curve is determined by certain aspects of the statistics of the magnetization fluctuations in the demagnetized state. Clearly, the microscopic details of the magnetization pattern in the demagnetized state depend on magnetostatic and exchange (if present) interactions between grains and microscopic disorder mechanisms such as random grain sizes and random granular anisotropy. Two very general models were considered by Bertram, Hallamasek and Madrid (1986) and their impact on the shape of the noise power vs. reverse write current curve assessed. If the noise is due to the intrinsic disorder of grains or of correlated clusters of grains then noise power along the major hysteresis loop will vary as $\{1-[M(I)/M_r]^2\}$. If, on the other hand, the noise is due to microscopic modulation of the recorded magnetization, such as would occur from micrometer-sized material inhomogeneities or fluctuations in coercivity, then noise power will vary as the squared derivative of the remanence magnetization loop, $(dM/dH)^2$. Silva and Bertram (1990) and Tarnopolsky, et. al. (1991), working in different alloy systems, measured the noise curves vs. reverse write current plus *in-situ* remanence magnetization curves via a long-wavelength recording spectra method for disks of varying H_c and $M_r\delta$ and compared the results with the two models. Both groups found the inhomogeneity model to fit the measured data very closely, that is, noise power varied as $(dM/dH)^2$.

However, other workers in the field have obtained different results from similar experiments and from micromagnetic modeling. Zhu and Bertram, for instance, found an excellent fit between experimental data and (dM/dH) rather than $(dM/dH)^2$. Micromagnetic modeling in the same paper of magnetization noise and remanence hysteresis loop properties also found that dc-demagnetized noise power is fit by (dM/dH) (Zhu and Bertram, 1991). Zhu discusses this disagreement at length in his chapter in this volume (Chapter 6). The discrepancy may be resolved by careful attention to experimental details and to demagnetizing fields present during the writing process, both in the in situ remanent hysteresis loop measurement and in the dc-demagnetized noise power measurement. The difference in functional form of the noise power vs. reverse write current curves is significant; it will tell whether the demagnetized noise (and probably transition noise) is caused by intrinsic disorder of grains or clusters of grains or by larger-than-grain-scale variations in media properties such as coercivity.

In additional work, both Silva and Bertram (1990) and Tarnopolsky, et. al. (1991) were able to derive approximate noise correlation lengths from the decorrelated noise power spectral densities and found that noise correlation length is a monotonically increasing function of noise power. That is (depending on which is cause and which is effect), higher noise power is associated with/caused by longer ranged fluctuations in magnetization in the medium. However, neither group has compared the spectral statistics of dc demagnetized noise and transition noise.

6.3 On-track ac-erased/demagnetized noise

If the dc demagnetized state exhibits characteristic noise then the ac demagnetized state will as well. As discussed earlier, in recording this state can be approximated by recording at a very high frequency as the disk spins under the head. Again, even though both these demagnetized states lie at the origin of the hysteresis loop, there is no *a priori* reason to expect them to be statistically identical since they are reached by significantly different magnetic histories. In fact, little work of a systematic nature has been done on the ac erased state. Some hints about the magnetic pattern to be expected are found in the paper by Arnoldussen and Tong (1986) where they show a Lorentz micrograph of a portion of a track recorded at very high frequency. Though not completely randomized, the magnetization pattern is broken into small islands of oppositely directed magnetization. The expectation is that an ac erased medium would be randomized into rather small regions, resulting in low noise power.

6.4 As-deposited virgin noise (similar to ac-demagnetized noise)

This state of the medium lies at or near the origin of the hysteresis loop, but, as discussed in Section 6.1, it is the naturally demagnetized state into which the medium falls during film growth. This type of noise may affect recording at the edges of a track and thus the noise experienced in a disk drive if the head reads back slightly off-track, unless the entire recording surface is first prepared in the very low noise dc erased state. Again, little work of a systematic nature has been done. Unpublished Master's thesis work by J. Hoinville, working under Professors M. Muller and R. Indeck at Washington University in St. Louis (1989) found for four disks that the virgin noise spectrum was intermediate in magnitude between dc erased noise and the spectra for maximum transition noise and dc demagnetized noise (which were similar to one another). The native magnetization pattern and statistics of this state may provide an additional tool for understanding something about the of the grain size, distribution and coupling mechanisms of the recording film.

6.5 Track edge noise

Not all the fluctuations of the magnetization in recorded media are confined to the center line of the recorded track. Fluctuations also occur along the edges of a recorded track and the nature of these fluctuations is significantly different from the fluctuations within a transition or in the case of dc-demagnetized noise. Yarmchuk

94

(1986) was the first to report on observations of this type of noise. Using a technique of digitizing and averaging the readback signal and subtracting signals to simulate different widths of readback heads, he found that the recording process induced a narrow band of noisy magnetization along the track edges. (Figure 11) He also found that the width of this band of noise seemed to be independent of track width. As a result, although it constitutes a small fraction of the noise contribution for wide tracks ($>10\,\mu$m), as track widths are reduced this noise would contribute an increasing portion of the total medium noise and eventually (at, perhaps, 1-2 μm head widths) most medium noise would come from the track edges; further track width reduction would not reduce medium noise any more.

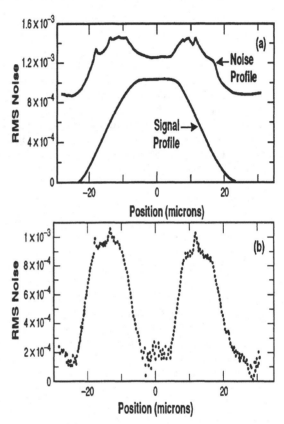

The character of track edge noise is significantly different from transition noise. Work by Muller, et. al. (1990) and Indeck, et. al. (1991b) has shown that track edge noise arises not from magnetization fluctuations caused by a head-on wall, like a transition, but from a different process. Assuming that a recording disk's surface has been prepared by dc-erasing the entire surface in one direction, as is typically done during drive manufacture, along the track edge the magnetization on- and off-track will be either parallel or anti-parallel for alternating bits. Even if the disk has been left in its virgin

Figure 11: Plot (a) shows the rms noise voltage vs. position as a 12.5 μm wide read is scanned across a 28 μm wide track. The noise is measured with a spectrum analyzer. For comparison, a plot of signal vs. position is also shown. Plot (b) shows the edge noise component obtained from the total by subtracting the dc erased and transition noise. (Amplitudes are arbitrary.) (Yarmchuk, 1986)

magnetization state before recording, the on-track magnetization will be rotating from its recorded direction to a demagnetized state off-track. In either case, there is no head-on magnetization change to accentuate fluctuations in the transition reversal zone. Thus, track edge fluctuations will be of lower magnitude than is found in the transitions for a given medium.

Moreover, the spectral distribution of edge noise differs from transition noise (Muller, et. al., 1990). A recording head which straddles a track edge is almost entirely sensitive only to the time rate of change of the x-component of magnetization (along the track), so the noise signal reflects changes in the location of the exact track edge. This can be seen by

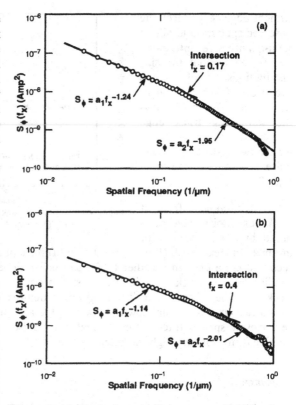

Figure 12: (a) Track edge noise flux spectrum of a high noise disk. (b) Track edge flux noise spectrum of a low noise disk. (Muller, et. al., 1990)

noting that a straight track edge under the head will produce no flux change in the head and so no signal, but if the location of the edge (defined where the magnetization is perpendicular to the track direction as the magnetization reverses through 180°) wanders then the flux into the head will vary, inducing a noise signal. The noise magnitude thus tells something about the spatial extent of the wandering edge and the Fourier spectrum of the track edge noise can be used to obtain further information about the nature of this wandering edge. This spectrum is easily obtained by decorrelating the noise signal by the read head's sensitivity function, as Muller, et. al. (1990) demonstrate. Figure 12 is an example of the resulting spectra for a high noise disk and a low noise disk. In both cases the noise spectrum follows an f_x^{-b} law but exhibits a kink, or slope change, at a particular wavenumber. Modeling of the track edge fluctuations in terms of uncorrelated excursions from a straight edge reproduces the observed spectra and allows semi-quantitative inferences to be drawn about the size distributions of the excursions from a straight track edge. Figure 13 shows the wandering track edge calculated from this model for the two disks whose edge noise spectra are shown in Fig. 12.

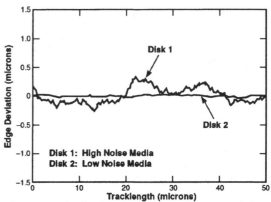

Figure 13: Simulated track edges, from the spectra of Figure 12. (Muller, et. al., 1990)

Additional effects of track edge noise have been noted in other work. For instance, Indeck, et. al. (Indeck, et. al., 1991a) have noted that when two tracks are recorded close enough together (≤ 0.5 μm or so, depending on specific recording conditions) their track edges interact in such a way that the noise at the edge of the track is significantly reduced. The magnetic structures formed between the tracks in such a case are, as yet, unknown. However, the abrupt reduction in edge noise for very closely spaced tracks suggests that edge noise may not turn out to be the limiting factor for very high track density recording.

7. Summary

Medium noise is of interest both because it affects the reliability of the data channel and because it is useful for understanding some properties of the recording medium itself and of the recording process that encodes data as a magnetization pattern. These related but distinct purposes lead to a variety of instruments and techniques for measuring and representing medium noise. For instance, if one desires the impact of medium noise on the bit error rate of a recording channel he might use a time interval analyzer to measure the uncertainty in timing of detected pulses. On the other hand, if one desires an understanding of the physical causes of medium noise then a spectrum analyzer is the most informative instrument. In this chapter we have reviewed many of the measurement methods in common use, although with an emphasis on the frequency domain techniques based on spectrum analyzer measurements. We have also shown that for gaining an understanding of the physical causes of medium noise it is necessary to compute the intrinsic noise power spectral density function from the raw noise spectrum, by removing the effects of the head transfer function.

Medium noise is caused by microscopic statistical fluctuations in the magnetization pattern and manifests itself in several different ways depending on the magnetic state of the recording medium. The type of noise of primary interest is transition noise. Along a recorded track the noise is significantly greater in the region where the magnetization reversal is occurring (the transition) than in the uniformly magnetized regions between transitions. This means that measurements whose purpose is the measurement of the effect of medium noise on channel error rate must measure the noisiness of the transitions themselves, not just the average noise along a stretch of recorded track. We discussed the characteristic properties of transition noise, including the extra noise which is observed for closely spaced transitions.

Other types of medium noise, corresponding to the magnetization fluctuations characteristic of other magnetization states of the recorded medium, were also discussed. It was shown that the statistical properties of these different types of noise differs (as reflected in their intrinsic noise power spectral densities), indicating different types of physical mechanisms causing the magnetization fluctuations. Reverse dc-demagnetized noise is particularly important since its magnetization state is very similar to that of the transition. Analysis of this noise may therefore shed light on the causes (and control) of transition noise. Similarly, the other types of medium noise all contribute to our understanding of the statistical processes which affect the regularity of the magnetization of the recorded bit. This understanding has already led to significant advances in reducing medium noise in thin film media; this trend can be expected to continue.

8. Acknowledgments

I wish to thank Lung Tran and Steve Naberhuis for technical assistance and Dick Baugh for helpful discussions during the preparation of this chapter.

9. References

(Anonymous) "Understanding and Measuring Phase Noise," Hewlett-Packard Application Note #AN-207, pg. 7 (October 1976)

Aoi, H., M. Saitoh, N. Nishiyama, R. Tsuchiya and T. Tamura, "Noise Characteristics in Longitudinal Thin-Film Media," *IEEE Trans. Mag.*, **22**, p. 895, (Sept. 1986)

98

Arnoldussen, T. and H. Tong, "Zigzag Transition Profiles, Noise, and Correlation Statistics in Highly Oriented Longitudinal Film Media," *IEEE Trans. Mag.*, **22**, p. 889, (Sept. 1986)

Barany, A. and H. N. Bertram, "Transition Noise Model for Longitudinal Thin Film Media," *IEEE Trans. Mag.*, **23**, p. 1776 , (March 1987)

Baugh, R. A., E. S. Murdock, and B. R. Natarajan, "Measurement of Noise in Magnetic Recording Media", *IEEE Trans. Mag.*, **19**, p. 1722, (Sept. 1983)

Belk, N., P. George, G. Mowry, "Noise in High Performance Thin-Film Longitudinal Magnetic Recording Media," *IEEE Trans. Mag.*, **21**, p. 1350, (Sept. 1985)

Bertram, H. N., K. Hallamasek and M. Madrid, "DC Modulation Noise in Thin Metallic Media and Its Application for Head Efficiency Measurements," *IEEE Trans. Mag.*, **22**, p. 247, (July 1986)

Bracewell, R. N., *The Fourier Transform and Its Applications*, (McGraw-Hill, 1978) pp. 25 and 46 and Chapter 10.

Che, X. and H. Suhl, "Magnetic Domain Patterns as Self-Organizing Critical Systems," *Phys. Rev. Letters*, **64**, p. 1670, (2 April 1990)

Fisher, R. D., L. Herte, A. Lang, "Recording Performance and Magnetic Characteristics of Sputtered Cobalt-Nickel-Tungsten Films," *IEEE Trans. Mag.*, **17**, p. 3190, (Nov. 1981)

Hoinville, J., "Noise in Longitudinal Thin Film Media," Master's Thesis, Washington University at St. Louis, (1989) unpublished

Indeck, R., S. Reising, J. Hoinville, and M. Muller, "DC Track Edge Interactions," *J. Appl. Phys.*, **69**, p. 4721 (1991a)

Indeck, R., G. Mian and M. Muller, "Determination of a Track's Edge by Differential Noise Power Spectrum," Paper MA-12, presented at 5th Joint MMM-Intermag Conference, Pittsburgh, PA, June 1991, to be published in *IEEE Trans. Mag.*, **27**, (1991b)

Katz, E., and T. Campbell, "Effect of Bitshift Distribution on Error Rate in Magnetic Recording," *IEEE Trans. Mag.*, **15**, p. 1050, (May 1979)

Madrid, M. and R. Wood, "Transition Noise in Thin Film Media," *IEEE Trans. Mag.*, **22**, p. 892, (Sept. 1986)

Majumdar, A. and B. Bhushan, "Role of Fractal Geometry in Roughness Characterization and Contact Mechanics of Surfaces," *ASME J. Tribology*, **112**, p. 205, (1990)

Muller, M. W., R. S. Indeck, E. S. Murdock, R. Ornes, "Track Edge Fluctuations," *J. Appl. Phys.*, **67**, p. 4683, (May 1990)

Murdock, E., B. Natarajan, R. Walmsley "Noise Properties of Multilayered Co-Alloy Magnetic Recording Media," *IEEE Trans. Mag.*, **26**, p. 2700, (Sept. 1990)

Ogawa, K. and S. Ogawa, "Noise Reduction of Magnetic Coated Disk," *IEEE Trans. Mag.*, **15**, p. 1555, (Nov. 1979)

Papoulis, A., "*Probability, Random Variables, and Stochastic Processes*," Ch. 9, 10, 14 (McGraw-Hill, 1965)

Silva, T., and H.N. Bertram, "Magnetization Fluctuations in Uniformly Magnetized Thin-Film Recording Media", *IEEE Trans. Mag.*, **26**, p. 3129, (Nov. 1990)

Tahara, Y., Y. Miura, Y. Ikeda, "Peak Shift Caused by Gaussian Noise in Digital Magnetic Recording," *Electr. and Commun. in Japan*, **59-C**, no. 10, p. 77, (1976)

Tanaka, H., H. Goto, N. Shiota, M. Yanigasawa, "Noise characteristics in plated Co-Ni-P film for high density recording medium," *J. Appl. Phys.*, **53**, p. 2576, (Mar. 1982)

Tang, Y. "Noise Autocorrelation in High Density Recording on Metal Film Disks," *IEEE Trans. Mag.*, **22**, p. 883, (Sept. 1986)

Tarnopolsky, G., L. Tran, A. Barany, H.N. Bertram and D. Bloomquist, "DC Modulation Noise and Demagnetizing Fields in Thin Metallic Media," *IEEE Trans. Mag.*, **25**, p. 3160, (July 1989)

Tarnopolsky, G., H. N. Bertram and L. Tran, "Magnetization Fluctuations and Characteristic Lengths for Sputtered CoP/Cr Thin Film Media," *J. Appl. Phys.*, **69**, p. 4730, (1991)

Terada, A., O. Ishii, S. Ohta, T. Nakagawa, "Signal-to-Noise Ratio Studies on γ-Fe_2O_3 Thin Film Recording Disks," *IEEE Trans. Mag.*, **19**, p. 7, (Jan. 1983)

Thurlings, L., "Statistical Analysis of Signal and Noise in Magnetic Recording," *IEEE Trans. Mag.*, **16**, p. 507, (May 1980)

Yarmchuk, E., "Spatial Structure of Media Noise in Film Disks," *IEEE Trans. Mag.*, **22**, p. 877, (Sept. 1986)

Zhu, J.-G. and H. N. Bertram, "Micromagnetic Studies of Thin Metallic Films," *J. Appl. Phys.*, **63**, p. 3248, (1988a)

Zhu, J.-G. and H. N. Bertram, "Recording and Transition Noise Simulations in Thin Film Media," *IEEE Trans. Mag.*, **24**, p. 2706, (1988b)

Zhu, J.-G. and H. N. Bertram, "Self-Organized Behavior in Thin Film Recording Media," *J. Appl. Phys.*, **69**, p. 4709 (1991)

Zhu, J.-G., "Noise of Interacting Transitions in Thin Film Recording Media," *IEEE Trans. Mag.*, **27**, (1991) in press

CHAPTER 4

THEORETICAL CONSIDERATIONS OF MEDIA NOISE

THOMAS C. ARNOLDUSSEN
IBM Corporation
San Jose, California 95193 USA

This chapter concentrates on the broad theoretical aspects of noise in recording media, such as statistical properties, rather than micromagnetic physics. However, when noise is the subject much of the physics *is* the statistics. A tutorial perspective is maintained by including sufficient references to illustrate specific points, without becoming a comprehensive literature review. Special attention is given to the implications of the spatial scale at which random fluctuations occur. In particular, this chapter addresses the distinctions between *additive* and *modulation* noise, which are scale-dependent manifestations.

1. Noise Terms, Descriptions, and Definitions

The study of information communication requires clear understanding of the concepts of information, signal, noise, distortion, and interference. Our interest here is noise, but we must define "what is not noise" in order to agree upon the meaning of experimental measurements, as well as to address the theoretical groundwork.

In radio communication, the information is usually an assembly of words, meaningful to the sender and receiver, which conveys a message. These words are transmitted by encoding them into an electrical representation (such as a varying voltage), which travels as electromagnetic waves to a receiving station. There the electrical variation is decoded (or detected), thus recreating the original audible sounds. The encoded electrical representation (a carrier wave modulation) is the signal. Upon decoding the signal, the words may not sound exactly the same as when the sender spoke them.

This change may be due to several causes. First, imperfect reproduction electronics (amplifier, speaker, etc.) may have a frequency response which changes the tones of the original spoken words. This is distortion, but unless it is extreme, the words are understandable and the message (information) is retrievable. Second, another radio station may be broadcasting near the same frequency. Due to a finite detector bandwidth the words of the second station could be intermingled with the primary broadcast. This is interference. Interference can and generally does contain information, but not that which is desired, and can confuse the intended message. Third, the incoming signal (and therefore the message) can become obscured by noise, characterized by randomness. Solar flares and random motions of electrons in resistors are two such sources of noise in radio communication.

Distortion is in principle a predictable alteration of the signal for a given transmitter and receiver. Interference, while not necessarily predictable in practice, is not random. It often contains real information and spectral content similar to the desired signal. Distortion and interference are not considered noise, though they can corrupt information. Noise is introduced physically in much the same way as distortion and interference, but distinguished by its **random** nature, devoid of information. One could describe noise as **random** distortion or interference, but by convention the terms distortion and interference are reserved for deterministic phenomena.

For digital data magnetic recording, the transmitted information is a meaningful sequence usually composed of the numbers "one" and "zero." As a simple example, every tick of a clock could generate a positive or negative voltage pulse for a "one" and a null voltage for a "zero." Thus a 0100101001001010 sequence could be encoded into the voltage waveform of Fig. 1, called the signal. Since the medium of "transmission" is the hard magnetic recording layer of a disk or tape, the voltage waveform of Fig. 1 triggers a current source which drives the recording head. When a positive voltage spike occurs the current through the recording head is changed from a negative current to a positive current. When a negative voltage spike occurs the current through the head reverses from positive to negative. When the voltage is zero, the current through the head remains as it was. Such current reversals or lack thereof in the recording head create the pattern of magnetization reversals in the medium, illustrated in Fig. 2.

Later when the reproduce head passes over this magnetization pattern magnetic flux coupled into the read head produces the string of voltage pulses shown in Fig. 3. Thus the signal has been "transmitted" from the recording electronics to the reading electronics. Note that although the voltage pulse sequence is the same in Figs. 1 and 3, and the information content is the same, the pulses have been given a different, broader shape in the process. The signal has undergone distortion, characteristic of the recording and reading head geometries, the head-to-medium spacing, and the magnetic properties of the disk or tape. Yet despite this or any other distortion (e.g., amplifier), Fig. 3 should be considered the signal (unless the goal is specifically to characterize distortion). If distortion is too great, errors may occur in the decoding process. The decoder usually requires a pulse to be above a minimum amplitude and its peak to be within a restricted timing window to register as a "one." If the disk coercivity is too low or the pulses are crowded too closely,

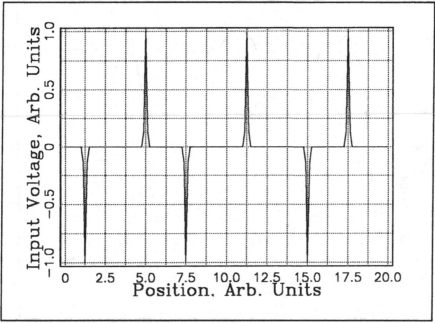

Figure 1. Signal input voltage pulses.

some pulses may have amplitudes below the clip level or have peaks shifted out of the timing window, and be erroneously decoded as "zeros" instead of "ones." Thus a signal could be free of noise and interference and yet have errors. The error rate depends on the distortion as well as the robustness of the encoding / decoding algorithm, and not simply on noise or interference.

If another set of data is written on an adjacent track, the read head will pick up that signal (interference) as well, through sidereading or simple mispositioning of the head. If a data pattern of 0101010001010100 were picked up 1/10th as strong as the principal track, the sum of the on-track signal and off-track interference might appear as in Fig. 3. Pulse amplitude variations and phase shifts are seen. A data pattern previously written on the track, but incompletely erased (or overwritten) when new data is recorded, can also be a source of interference. Obviously such interference can produce errors upon decoding. With a waveform so similar to the signal of interest, the potential for corrupting data can be greater than (random) noise. Returning to the radio broadcast analogy, it is often easier to understand a broadcast of speech in which considerable noise is present than one with which a competing station interferes. Moreover, if the interfering station broadcasts in the same language, the confusion is much greater than if it broadcasts in a language unknown to the listener.

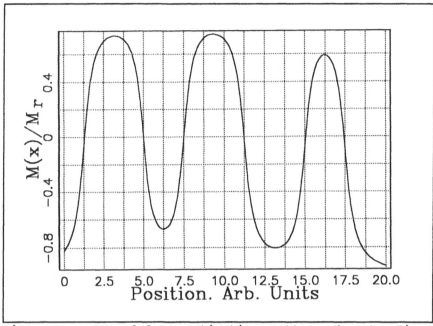

Figure 2. Recorded magnetization pattern due to Fig.1 input.

1.1 Definitions

Information is the meaning or message to be communicated.

Signal is the (electric/magnetic) representation of the information being communicated, free of noise and interference, but not necessarily free of distortion.

Distortion is any deterministic change of the signal shape due to the characteristics of the recording, storage, and readback system. It can be linear or nonlinear. Distortion is only a meaningful term when comparing "similar" signal representations of the information. The broadening of readback voltage pulses compared with write-driver voltage pulses can meaningfully be termed distortion. However, the translation of write-driver voltage pulses into the storage format of magnetization reversals should not be considered distortion, but rather signal transformation.

Interference is the introduction of an unwanted deterministic signal into the desired signal. If introduced only in the readback (as for adjacent track pickup), it

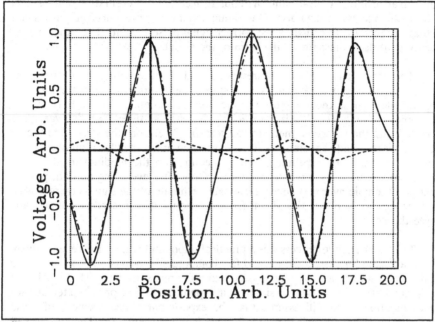

Figure 3. Readback voltage sequence corresponding to input signal of Fig.1. Solid curve is ontrack signal; dashed curve is adjacent track pickup; dash-dot curve is the sum.

simply adds to the desired signal. If the unwanted signal interacts with the recording of the desired signal (as may happen in overwriting old data with new data) it can be multiplicative, i.e., it can modulate the desired signal.

Noise is the random component of the electrical output of the readback head and associated electronics in the presence of a signal (with or without distortion) and possibly in the presence of interference. Quite simply it is what remains when signal and interference have been subtracted from the output waveform. This is the most pertinent definition of noise, in terms of the output of the read head and amplifier. However, it is useful to generalize the definition to deal with noise as it exists in the recording medium. **Media noise** is the random component of the magnetization pattern produced when a signal is recorded, or any spatially random property of the medium (such as surface roughness), which gives rise to noise in the electrical output of the read head and electronics.

As simple and intuitive as this noise definition appears, it is extremely difficult to measure (or theoretically treat) noise except for a few simple conditions. Even in the absence of interference, it is laborious at best to determine precisely what is "signal" in an arbitrary data stream. Without an accurate representation of the signal, it is impossible to assess the noise. One cannot assume any preconceived pulse shape (such as Lorentzian) to construct the correct signal pulse stream. The experimental measurement itself must reveal the form of the signal.

The pedagogical noise measurement would require writing the desired data pattern and reading it back many times, storing each timebased record, e.g., in a computer file. These records must then be averaged to eliminate the time varying (electronic) noise. Yet this is insufficient, because statistical spatial variations associated with the recording medium still remain. Therefore the same data record should be written and repeatedly read and averaged at another location on the disk or tape. This must be repeated at many locations on the medium to ultimately average out the spatially random variations. Only after averaging all data pattern averages (ensemble average) does a reasonable estimate of the signal emerge. And only then can we subtract this signal from individual data records to obtain and analyze the noise.

This cumbersome process was a practical impossibility until the recent advent of high rate sampling oscilloscopes and fast computers with massive storage. Even so, this is an impractical everyday approach to noise measurement. The solution to this dilemma has been to use very simple "data" patterns. Three principle techniques deserve mention. We will assume that the experimental measurement of noise always can be designed such that interference is not present. (a) The simplest of these techniques is the dc-erase. The medium is passed through a magnetic field from a recording head or erase magnet, which may have a spatial variation but does not vary explicitly with time. Often the peak field is chosen large enough to saturate the medium, leaving it in the "saturation remanent" state. Other times, after the medium has been saturated, a reverse lower strength field is applied to leave the medium in a non-saturation remanent state, such as the zero average magnetization "coercive remanent state." (b) Another approach is to bulk demagnetize the medium by applying a field which reverses periodically while the amplitude of the applied field is slowly decreased to zero. The field either should be spatially broad or should reverse direction at a low frequency, such that each particle or grain of the recording medium experiences many field reversals of decreasing magnitude as the disk rotates under the erase field. This is termed "bulk ac-erasure." (c) Yet another method is to write an "all-ones" pattern where the recording head reverses its field at a constant repetition rate typical of data recording frequencies, causing the magnetization of the medium to reverse with the same periodicity. In the first two cases, the form of the "signal" is simple: there is no signal (an average dc magnetization state, whether zero or nonzero, produces no voltage in the readback head). Thus anything read back by the head must be noise. Nevertheless dc and ac erasure leave the medium in different micromagnetic states with different noise content. In the third case, the exact time or spatial variation of the signal is neither determined nor necessary if one is content to measure rms total noise or rms spectral noise distribution. Using a spectrum analyzer to measure the rms noise spectrum, one ignores spectral peaks which are harmonics of the "all-ones" frequency written. The rest is noise.

The strength of all three approaches to noise measurement is simplicity. However, spectrum analyzer rms noise measurements sacrifice all knowledge about spatial or temporal noise distribution. Moreover, the noise can depend on the signal written (for thin film media this is generally the case, less so for particulate media), and simple signal patterns give only partial insight into how noise corrupts an actual data signal. Because these three conditions of noise measurement are both prevalent and pragmatic, they encompass the majority of noise publications and will receive most attention here. Short of the pedagogical approach to noise and signal measurement described above, rms measurements are the only practical methods for separating medium and head/electronic noise contributions. From a theoretical perspective, however, a discussion of spatial and temporal noise distributions will be included in section 3.

1.2 Additive and Multiplicative Noise

Before mathematically describing noise, it is useful to point out two important forms in which noise may arise, with attention directed specifically at media noise. One form is termed "additive" noise and the other "multiplicative" or "modulation" noise. In Section 2.3 the distinction between additive and multiplicative noise is elaborated.

1.2.1 Additive Noise.

Additive noise, as the adjective suggests, is noise which adds to the signal. However this could be said of any form of noise in the sense that noise is the difference between a (voltage) record and the signal. The distinctive aspect of additive noise is that it is functionally unrelated to the signal. A time record of noise voltage bears no functional similarity to the signal over the same time interval. Consider that a noise voltage record, at some time t, is decomposed into a number of noise events which occur simultaneously within an incremental time Δt. For example, individual magnetic particles distributed across the width of a recorded track, with roughly the same position along the track, induce simultaneous voltages in the read head, which outputs their sum. If these noise events are statistically independent of each other and their expected sum is zero, the noise is called *additive*. However, the mean square value of these simultaneous noise events, in general, will be nonzero. In fact, these mean square values may be distributed in time in a way that *is* functionally related to the signal time record, as for transition noise in film media, and yet be additive. In terms of the medium, this type of noise arises from microscopic randomness in magnetization or other properties. This might result from grain-to-grain variations in a film medium or a random distribution of positions of particles in a particulate medium.

1.2.2 Multiplicative Noise.

Noise which arises from macroscopic, or mesoscopic, randomness of the recording medium is proportional to the signal or a derivative of the signal. Again, if the noise record at time t is decomposed into a number of noise events which are simultaneous within a time increment Δt, and if these events are correlated, their expected sum is, in general, non-zero. So is their mean square value. Such noise is called *multiplicative*. For example, the recorded signal amplitude is proportional to the remanent magnetization. If the magnetic remanence varies randomly on a scale comparable to or greater than a bit cell (mesoscopic), such that the local average signal is affected, a random component is introduced which is proportional to the signal itself. Such proportionality implies that the randomness multiplies the signal or its derivative. Such multiplication is also called signal *modulation*. In the frequency domain, this is characterized by noise clustered about the harmonics of the signal, that is, modulation sidebands.

2. Mathematical Formulation of Media Noise

During readback, the voltage record observed after the read head and preamplifier can be decomposed conceptually into signal, media noise, and head/electronics noise. The head/electronics noise is an explicit random function of time. On the other hand, the signal and media noise can be considered implicit functions of time and explicit functions of position of the medium relative to the read head by the relation $t = t_0 + x/v$, where x is position along a recording track and v is the disk velocity. Here we are interested only in the signal and contributions to noise from the media. Therefore our formulation will be in terms of spatial distributions rather than temporal records. Designating the signal voltage as s(x), the media noise voltage as n(x), and the sum of these as e(x), the readback voltage record can be written

$$e(x) = s(x) + n(x)$$
$$= -N_c \epsilon \frac{d\varphi}{dt} = -N_c \epsilon v \frac{d\varphi}{dx} \tag{1}$$

where N_c is the number of coil windings on the read head, ϵ is the head efficiency, and φ is the magnetic flux from the medium which is coupled into the head. By reciprocity the readback voltage is given by Eq.(2), where $h(x,y,z)$ is the head sensitivity

$$e(x) = -N_c \varepsilon v \frac{\partial}{\partial x} \int\limits_{-\infty}^{\infty} dz' \int\limits_{d}^{d+\delta} dy' \int\limits_{-\infty}^{\infty} dx' \; \mathbf{h}(x',y',z') \cdot [\mu_o \mathbf{m}(x'-x,y',z')] \quad (2)$$

function, i.e., the vector field produced by the head normalized by the amp-turns applied; m(x,y,z) is the vector magnetization pattern in the medium; x is position along the recording track; y is position normal to the medium plane measured from the poletips of the head; and z is position in the direction transverse to the track. "d" is the distance from the head to the top surface of the medium. δ is the thickness of the medium. The readback voltage of Eq.(2) can also be written as (Heim and Monson, 1987)

$$e(x) = -N_c \varepsilon v \int\limits_{-\infty}^{\infty} dz' \int\limits_{d}^{d+\delta} dy' \int\limits_{-\infty}^{\infty} dx' \; h_x(x',y',z') \; [\mu_o \rho (x'-x,y',z')] \quad (3)$$

where h_x is the x component of h and $\rho(x,y,z)$ is the magnetic charge density equal to $-\nabla \cdot$ m. Both forms are useful, but we shall focus on Eq.(2) which leads to analytical expressions for media noise similar to treatments in the literature over the past 25 years. Eq.(3) is more convenient for dealing with media which can sustain two or three orthogonal components of magnetization, such as isotropic media and narrow track recording conditions.

There are two forms in which we want to calculate and, if possible, measure noise. First is the expected noise-squared distribution $<[n(x)]^2>$ described earlier as a pedagogical approach. To measure this noise distribution, the signal must be known so it can be subtracted from each voltage record of the ensemble. This is an impractical routine measurement, but can be a useful formulation for theoretical studies, and may have strong implications for choice of signal detection scheme. The second formulation relates to practical measurements, and that is the spectral distribution of noise-squared, e.g., volts-squared per Hz in the temporal frequency domain or volts-squared per wave-vector in the spatial frequency domain. Because voltage squared is proportional to power in passive circuits, the term "noise power spectrum" or NPS has come to be used. This can be expressed as in Eq.(4), where

$$NPS(k) = \lim_{L\to\infty} \frac{1}{L} [E_L(k) \; E_L^*(k) \; - \; S_L(k) \; S_L^*(k)]$$

$$= \lim_{L\to\infty} \frac{1}{L} N_L(k) \; N_L^*(k) \quad (4)$$

$E_L(k)$, $S_L(k)$, and $N_L(k)$ denote the Fourier transforms of $e(x)$, $s(x)$, and $n(x)$ and the approximate transform integration is performed over an x domain of length L which approaches infinity in the limit. Throughout this chapter, lower case letters denote functions of x and upper case the Fourier transforms with respect to x. A * indicates a complex conjugate. k is a wavevector in the x direction, with $k = 2\pi/\lambda$. In real space the quantity corresponding to the NPS is termed the autocorrelation function, acf(x), the inverse Fourier transform of NPS(k), and is given by:

$$acf(x) = \lim_{L \to \infty} \frac{1}{L} \int_{-L/2}^{L/2} dx' [e(x')e(x'+x) - s(x')s(x'+x)]$$
$$= \lim_{L \to \infty} \frac{1}{L} \int_{-L/2}^{L/2} dx' \, n(x')n(x'+x) \qquad (5)$$

The acf(x=0) is the mean squared noise voltage and is identically equal to the integral of NPS(k) over all $k/2\pi$. Equations (4) and (5) are general, assuming only that the (finite) limits exist as $L \to \infty$, and that n(x) is random (which it is by definition). Sometimes stationarity is stipulated as a requirement, not because these equations are invalid for non-stationary noise, but for reasons of interpretation, as discussed further in Section 2.3.1 of this chapter.

2.1 Particulate Additive Noise Formulation

We will now go through a simple, but instructive, calculation of noise, using a set of assumptions which were fairly realistic for particulate media of the past (Mee, 1964; Smaller, 1965; Mallinson, 1969; Daniel, 1972). Namely, let us assume the medium is a coating of thickness δ composed of non-interacting, identical magnetic particles. Assume the positions of the particles in the coating are statistically uncorrelated. (Since particles cannot overlap, this implies a low volume packing fraction, not strictly true of real coatings where 20 - 40 % of the coating volume is particles. In reality, particles do interact, they are never identical, and most importantly, they are often spatially correlated to some extent due to high packing fractions, clumping, and chaining in non-ideal dispersion processes.) For a bulk ac erasure, the following treatment does not depend on the assumption of low packing fractions because random particle polarities ensure uncorrelated particles. For arbitrary dc magnetization, we will show later that noise should depend on the average magnetization, but for very low packing fractions this can be neglected. The magnetization in Eq.(2) then represents an array of randomly distributed particles which pass by the read head without correlation to one another. The readback voltage is completely random and is therefore noise. Any dc component in the coating magnetization produces zero voltage in the read head. For relatively wide tracks, it is also customary to ignore track edge effects and to assume that the head sensitivity function is only a function of x and y. Furthermore, tapes and rigid disks employ coatings with the acicular particles oriented along the track direction. Hence, it has also been customary to assume only an x-component to magnetization, so that only h_x and $m_x = m$ appear in Eq.(2). For the above assumptions

$$NPS(k) = \lim_{L \to \infty} \frac{1}{L} (N_c e v \mu_o)^2 k^2 \times$$

$$\int_{-W/2}^{W/2} dz \int_{-W/2}^{W/2} dz' \int_d^{d+\delta} dy \int_d^{d+\delta} dy' H_x(k,y) H_x^*(k,y') \times$$

$$M^*(k,y,z) M(k,y',z') \qquad (6)$$

dropping the "L" subscript from the Fourier transforms, but implying it. We may rewrite M(k,y,z) in the discrete notation $M(k,y,z) = \{M_p(r_p,k,y-y_p,z-z_p)\}$, to mean that M consists of an array of discrete particles whose centers are located by the vector $r_p = (x_p,y_p,z_p)$. If the particles have length x_c, thickness y_c, and width z_c then each m_p becomes zero for $|x-x_p| > x_c/2$ or $|y-y_p| > y_c/2$ or $|z-z_p| > z_c/2$. Within a particle, we assume the magnetization is uniform and equal to $\pm m_o$. NPS(k) can then be written:

$$NPS(k) = \lim_{L \to \infty} \frac{1}{L} (N_c e v \mu_o)^2 k^2 \times$$

$$\sum_p \sum_q \int_{-W/2}^{W/2} dz \int_{-W/2}^{W/2} dz' \int_d^{d+\delta} dy \int_d^{d+\delta} dy' H_x(k,y) H_x^*(k,y') \times$$

$$M_p^*(r_p,k,y-y_p,z-z_p) M_q(r_q,k,y'-y_q,z'-z_q) \qquad (7)$$

The integrations of Eq.(7) need only be performed over the particle volumes, where magnetization is nonzero. The Fourier transform of each particle contains the factors $\{x_c[\sin(kx_c/2)]/[kx_c/2]\}\exp(-ikx_p)$. When the result is summed over all particle locations, r_p and r_q only the p=q terms may be expected to contribute, due to the random phase factor $\exp(ik[x_p-x_q])$, for low packing fractions. For ac erasure, random particle polarities ensure that only p=q terms contribute, even for high packing fractions. Assuming that h_x is constant within a particle, we evaluate its transform as $H_x(k,y_p)$. NPS(k) then becomes:

$$NPS(k) = \lim_{L \to \infty} \frac{(N_c e v \mu_o)^2 k^2}{L} \times$$

$$\sum_{all\,p} |H_x(k,y_p)|^2 (m_o x_{cp} y_{cp} z_{cp})^2 sinc^2(kx_{cp}/2) \qquad (8)$$

Under the above assumptions x_{cp}, y_{cp}, and z_{cp} are the same for all particles and the "p" subscripts are shown only to emphasize that the summation is over all particles in the volume $WL\delta$. The summation can be transformed into an integration over y by multiplying by the average number of particles in a lamina of thickness dy ($N_v WLdy$, where N_v is the average number of particles per unit volume) and noting that every lamina is statistically the same. Thus

$$NPS(k) = \lim_{L \to \infty} \frac{(N_c ev\mu_o)^2 k^2}{L} \int_d^{d+\delta} dy |H_x(k,y)|^2 \times$$
$$[m_o x_c y_c z_c sinc(kx_c/2)]^2 [N_v WL] \qquad (9)$$

2-D head fields obey the functional relationship $\mathbf{H}(k,y) = \mathbf{H}(k,d)\, e^{-|k|(y-d)}$, for which the integration in Eq.(9) gives:

$$NPS(k) = (N_c ev\mu_o)^2\, k^2\, W\delta\, |H_x(k,d)|^2 \left(\frac{1 - e^{-2|k|\delta}}{2|k|\delta} \right) \times$$
$$N_v\, (m_o x_c y_c z_c\, sinc(kx_c/2))^2 \qquad (10)$$

If the medium were only one lamina thick with magnetization at any position (x,z) uniform throughout its thickness (as for grains in a thin film) the integration over y would have given the factor $[(1-exp(-|k|\delta))/|k|\delta]^2$.

The noise expression given by Eq.(10) is similar in form and derivation to those appearing in the literature, especially prior to the late 1980's (Mee, 1964; Smaller, 1965; Mallinson, 1969; Daniel, 1972). Recording media were virtually all particulate and therefore the focus was on the particle statistics, so much so that some treatments begin with or deal exclusively with statistics of magnetization fluctuations in the medium. Despite all these simplifying assumptions, Eq.(10) does give valuable insight into some major aspects of additive media noise. For example, the additive noise power is proportional to particle density N_v, and to the square of the particle volume for a constant m_o. For short particles or long wavelengths ($|k|x_c/2 << 1$), the noise power spectrum attributable to the media is nearly white, with the voltage noise power spectral shape being attributable to the spectral sensitivity of the head, the spacing loss, the readback differentiation process, and the coating thickness effect. While the signal power would be expected to vary as the track width squared, the noise power is proportional to W to the first power. These results provided key insight that drove the technology to use smaller and smaller magnetic particles to achieve higher signal-to-noise ratios. They also made clear the signal-to-noise cost of reducing trackwidths.

Nevertheless, this simple noise formulation fell short of predicting observed noise amplitudes and spectra for real particulate media, and the assumptions were inappropriate for films. Thurlings (1980) improved upon this model by striving to handle the statistical analysis of an array of particles properly, including longitudinal and perpendicular components of magnetization, as well as localized particle correlations. In this work, discrete magnetic particles were again the focus of the analysis, but the rather unphysical assumption was made that the particles had finite lengths and zero cross-sectional area. This allowed retention of the assumption that particles were randomly positioned. Particle volume fluctuations were handled by assigning dipole moments to the particles and allowing for a statistical distribution of these moments, which implicitly are the products of particle magnetization and volume. By letting particle positions be completely random and uncorrelated, or even clustered, this treatment permitted situations where real finite volume particles would overlap if they occupied the same positions assumed in the model. This mathematical convenience of "zero volume" particles was employed at the expense of violating physical reality. Whether this is a serious limitation of the model is arguable, but Thurlings found significant discrepancy between measured and calculated noise for a dc and ac erased particulate medium.

A more recent noise analysis (Nunnelley et al., 1987) achieved considerable generality, as well as agreement with experimentally measured noise, by avoiding *a priori* modeling of micromagnetic correlations directly. Equation (11) is their noise formulation. On the surface this expression is similar to Eq.(10), but the interpretation fundamentally distinct. This formula was derived for particulate media specifically, but the approach and interpretation is easily adapted to film media. As

$$NPS(k) = (N_c e \mu_o v)^2 W \delta |H_x(k, d)|^2 \left(\frac{1 - e^{-2|k|\delta}}{2|k|\delta} \right) k^2 \times$$
$$N_v \langle m^2 a^2 |F(k, l)|^2 \rangle \tag{11}$$

in most other noise treatments, the magnetization was assumed to be along the track direction with perpendicular or transverse components ignored. The major innovation in this work was to avoid addressing particles, as such, in analyzing the statistics, and instead introduce the concept of "independent noise sources." Rather than using physical dimensions of particles, the authors defined the size aspects of these noise sources in terms of autocorrelation lengths (along the track direction) and autocorrelation cross-sectional areas. Thus a cluster of particles (or grains of a thin film) whose magnetizations are spatially correlated are treated correctly as one noise source. The details of the nature and origin of such correlation within the noise unit is left for separate, independent modeling such as Monte Carlo simulations (Arratia and Bertram, 1984). The Nunnelley et al. analysis did not address the micromagnetic or mechanical origins of particle correlations but described the proper statistical treatment of noise sources and serves as a useful framework to extract correlation lengths and areas from measured noise.

Referring to Eq.(11), the factors on the first line are similar to those found in Eq.(10), and not directly related to the intrinsic noisiness of the medium. It is the second line which has a modified interpretation. N_v, is the average number of **noise sources** per unit volume, not the particle density. N_v as well as the other parameters are not fixed for a given medium, but depend on the state of "recording" which has been performed. A bulk ac demagnetization, a dc saturation, a dc full or partial demagnetization, or a periodic signal can all, in principle, produce different numbers and sizes of noise sources. The brackets $<...>$ imply the parameters within are averaged over all noise sources. m^2 is the mean squared magnetization within a single noise source, "a" is the cross-sectional area of the same noise unit as defined by autocorrelation transverse to the track direction, and "l" is the autocorrelation length of the source in the track direction. The function F(k,l) is the Fourier transform of the noise unit's geometry in the track direction. As for particles, if the noise unit is taken to be uniform in magnetization and cross-section along its length "l," then F(k,l) = [sin(kl/2)]/(k/2). The quantity on the second line of Eq.(11) can thus be interpreted as the (spectral) average squared dipole moment per unit volume with SI unit dimensions A^2m.

One cautionary note should be made. The magnetization appearing in the second line of Eq.(11) should be understood as the actual magnetization minus the value of magnetization corresponding to the signal (ensemble average over all tracks at the same x-coordinate as the noise source). For an average dc magnetization case, the dc average may be considered the signal for consistency of definition. Nevertheless, when calculating the voltage NPS, as above, it is unnecessary to subtract the average dc magnetization because it induces no voltage in the read head. However, the form of the NPS expression shown above invites one to divide Eqs.(11) or (10) by all the factors on the first line not directly related to noisiness of the medium, and call the remaining factors on the second line the "flux noise power spectrum per unit volume of medium," analogous to the term "power" for the voltage squared noise spectrum. Nunnelley et al. define this as S(k), but to avoid confusion with signal it will be denoted here as FNPS. When calculating FNPS average signal magnetization values must be subtracted even if they are dc, because the differentiating factor (k^2) has been removed. If the volumetric packing fraction of particles or grains is less than 100%, and the signal magnetization has been subtracted, the averaging must be performed over both the particles and spaces between particles. Thus

$$FNPS(k) = N_v \left\langle (m-\langle m \rangle)^2 \; a^2 l^2 \; \frac{|F(k,l)|^2}{l^2} \right\rangle \tag{12}$$

$<m>$ denotes the signal (average magnetization if magnetized to a uniform level).

Equation (12) defines the proper averaging that must be done in calculating the noise. For example, if "l" and "a" are correlated, as for hypothetical ellipsoidal

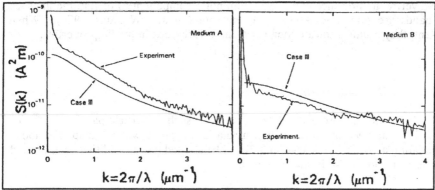

Figure 4. Flux noise power spectrum for two particulate disk media. Medium A particles are about 5.5 times the volume of medium B. (© IEEE; Nunnelley, et al., 1987)

particles of constant shape but distributed size, the noise will be proportional to $<l^6>$, whereas for cylindrical noise sources with lengths uncorrelated to cross-sections the noise varies as $<a^2><l^2>$. The former will weight large volume particles more heavily than the latter. Figure 4 shows a comparison between measured and modeled noise for two particulate media (Nunnelley et al., 1987). They explored possible correlations of particle cross-sections and lengths (assuming ellipsoidal and cylindrical shapes) and demonstrated that the noise spectral shape and magnitude could not be explained in terms of individual particles. Cylindrical particles with lengths uncorrelated to cross-sections were not only closer to the observed microscopic particle shape distributions, but gave closer agreement with the magnitude of the measured noise spectrum. Only by assuming a statistical probability that particles chain together could they match both the spectral shapes and amplitudes, concluding that in these media average chain lengths were about four particles long.

Below $k \sim 0.2\ \mu m$ the noise spectra of Fig.4 are not accurately described by this model. Two possible explanations are given for the sharp noise peak near $k = 0$. First, the experimental FNPS was obtained by dividing the experimental NPS by the calculated spectral characteristic of the head and the differentiating factor k^2. This involves dividing a measured quantity approaching zero (at small k) by a calculated value also approaching zero, which can result in significant error. Second, the apparent sharp noise peak near $k = 0$ could be due to a component which is true modulation noise. This portion of the spectrum was not explored in the Nunnelley et al. paper. We only draw attention to this feature to point out that this portion of the experimental FNPS is subject to large errors, but may indicate two simultaneous noise contributions, one being additive and the other being a modulation effect. Observing spectra in the presence of an "all ones" signal often reveals similar narrow sidebands of noise near the signal harmonics for particulate media. Such a test would verify the presence of a modulation noise component. For film media, such sidebands are sometimes seen, sometimes not (Su and Williams, 1974). When present, they usually indicate long wavelength variations in media properties.

2.2 Film Media Additive Noise

We could explore particulate media noise such as particle packing effects or clustering, but since our focus is film media in this book, we will adapt the above formulation accordingly. Equation (12) will be applied first to additive dc-erase and then to transition noise in thin film media.

2.2.1 DC-Erase Noise

One popular measurement is to saturate the medium in one direction, partially or fully dc erase with a field in the opposite direction, and then measure the noise level. We can analyze this measurement as follows. First assume the medium has virtually a 100 % volume packing fraction. Second, assume the medium magnetization easy axis is oriented along the track direction (for dc demagnetization noise the final result will still be a good approximation even for grains with randomly oriented easy axes). Finally, assume that for a perfect bulk ac erasure with a unidirectional field, the medium is broken up into minimum sized domains or clusters with average correlation lengths along and transverse to the track direction of l_c and w_c. Consider the film magnetization to be uniform throughout its thickness δ. We can treat the medium as if it consisted of strips in the x direction of width w_c, each strip uncorrelated with the rest. Absence of correlation in the z-direction can be assumed even when regions of adjacent strips have the same magnetization polarity by taking the strips to be randomly phased in the x-direction. If w_c is much less than the read trackwidth, the noise will be additive despite the dense spatial packing of noise sources. Figure 5 illustrates the model, where light and dark blocks imply opposite polarity magnetizations. l_c and w_c are depicted as constant in the figure, but should be understood to have a statistical variation.

If the medium is dc demagnetized to an average value $<m>$, with each l_c by w_c unit having either $\pm m_o$ magnetization, the probability that a given unit is magnetized in the positive direction is p and in the reverse direction (1-p). The average magnetization is $<m> = m_o (2p - 1)$. Within a given strip will be randomly distributed chain lengths of contiguous magnetic units having the same polarity. These chains will constitute the noise units with equal numbers of positive and negative polarity chains, regardless of $<m>$. Simple statistics reveal that the average length of positive and negative chains is:

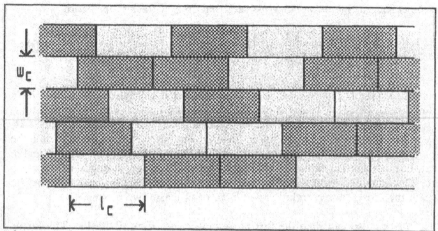

Figure 5. Model of film media dc magnetization state consists of strips of correlation width w_c having subunits of correlation length l_c long. Phase of strips is random.

$$\langle l_+\rangle = \frac{\langle l_c\rangle}{1-p} \ , \ \ \langle l_-\rangle = \frac{\langle l_c\rangle}{p}$$

$$and \ \ \langle l\rangle = \frac{1}{2}[\langle l_+\rangle + \langle l_-\rangle] = \frac{\langle l_c\rangle}{2\,p\,(1-p)} \tag{13}$$

In Eq.(12) $N_v = [(L/\langle l\rangle)(W/\langle w_c\rangle)]/(LW\delta)$, which (upon writing p in terms of $\langle m\rangle/m_o$) is

$$N_v = \frac{\left[1 - \left(\frac{\langle m\rangle}{m_o}\right)^2\right]}{2\,\langle w_c\rangle\,\langle l_c\rangle\,\delta} \tag{14}$$

The total mean square flux noise per unit volume of medium is given by integrating Eq.(12) with respect to $k/2\pi$ from minus to plus infinity (wide bandwidth). Taking F/l to be sinc(kl/2), the integral over all $k/2\pi$ of the bracketed factor in Eq.(12) becomes

$$\langle \delta^2 w_c^2(m - \langle m\rangle)^2 l\rangle = \frac{1}{2}\delta^2\langle w_c^2\rangle[(m_o - \langle m\rangle)^2 \langle l_+\rangle + (m_o + \langle m\rangle)^2 \langle l_-\rangle]$$

$$= 2\,\delta^2\langle w_c^2\rangle\langle l_c\rangle\,m_o^2 \tag{15}$$

The mean square flux noise per cross-sectional area of track is therefore

$$\frac{1}{2\pi}\int_{-\infty}^{\infty} dk \; FNPS(k) \; = \; \frac{\langle w_c^2 \rangle \, \delta}{\langle w_c \rangle} \; m_o^2 \left[1 \, - \, \left(\frac{\langle m \rangle}{m_o} \right)^2 \right] \tag{16}$$

The units of noise in this result may be more intuitive if multiplied by $(\mu_o^2 W\delta)$ from the first line of Eq.(11). Then the expression would be mean square noise with units Webers squared, where Webers are recognized as a unit of magnetic flux. By convention, free space permeability is factored out since it is a constant and the cross-sectional area $W\delta$ removed to normalize Eq.(12). Equation (14) represents the density of noise sources, while Eq.(15) represents the average "strength" of the noise sources. Combined with experimental measurements, Eq.(16) can be used to estimate the cross-track correlation width w_c, or at least $\langle w_c^2 \rangle / \langle w_c \rangle$.

We can also calculate the flux noise power spectrum, FNPS(k). The factor $\sin^2(kl/2)$ must be averaged over all chain lengths. The probability that a positive chain has length $l=nl_c$ is $(1-p)p^{n-1}$, and the probability that a negative chain has length $l=nl_c$ is $p(1-p)^{n-1}$. The FNPS(k), written in terms of $p = (1+(\langle m \rangle/m_o)^2)/2$ and $q = 1-p$ for conciseness, is

$$FNPS(k) \; = \; \frac{8pq\delta \langle w_c^2 \rangle \langle l_c^2 \rangle m_o^2}{\langle w_c \rangle \langle l_c \rangle} \; \times$$
$$\left[\frac{q^3}{p} \frac{1}{\ln\!\left(\frac{1}{p}\right)\!\left[\left(\ln\!\left(\frac{1}{p}\right)\right)^2 + (kl_c)^2\right]} \; + \; \frac{p^3}{q} \frac{1}{\ln\!\left(\frac{1}{q}\right)\!\left[\left(\ln\!\left(\frac{1}{q}\right)\right)^2 + (kl_c)^2\right]} \right] \tag{17}$$

For complete erasure, $\langle m \rangle = 0$ and $p = q = 1/2$. The FNPS(k) has a pure Lorentzian shape in k, i.e., $1/[1 + (kl)^2]$. This very form has been reported in the literature for several media (Tarnopolsky et al., 1991). In Chapter 3, FNPS(k) $\propto$ 1/k is reported, which may be due to non-binomial statistics being operative or due to viewing data over a limited range of k.

The prediction that dc noise should be proportional to $[1-(\langle m \rangle/m_o)^2]$ can be derived in a much simpler manner (Bertram and Silva, 1990) by noting that the variance of magnetization, obeying binomial statistics, is proportional to $p(1-p) = .25[1-(\langle m \rangle/m_o)^2]$. The more laborious approach above illustrates that the strength of the noise sources is not changing with average magnetization, but rather the number. A secondary change in the strength of the noise sources can occur (if w_c and/or l_c have a slight dependence on $\langle m \rangle$), but the change in number with average magnetization remains the primary effect (see Chapter 6). Bertram and Silva (1990) have reported that dc-erase noise varies as $(dM/dH)^2$, rather than $[1-(\langle m \rangle/m_o)^2]$, implying modulation rather than additive noise. Chapter 6 discusses this in detail, showing by experiment and modeling that the noise actually varies as (dM/dH), which is approximately the same as $[1-(<m>/m_o)^2]$. The conclusion is that dc-erase noise is primarily additive in nature, not modulation.

2.2.2 Transition Noise

Equation (12) can also be used to give an approximate expression for "transition noise" in film media. This is noise localized in the vicinity of a written magnetization reversal and, in fact, is noise which is created by the very act of writing a reversal. Chapters 5 and 6 show example images of recorded magnetization transitions (calculated or experimentally obtained) which illustrate how average magnetization reverses across a transition boundary. Significant microscopic irregularities are seen in the transition region. Such irregularities are the principle source of noise in thin film media.

If an "all ones" signal (equally spaced, alternating sign pulses) is written, there will be two transitions per wavelength λ (signal fundamental). The number of noise sources in a transition of track width W will be $W/<w_c>$, where $<w_c>$ is the average cross-track correlation width, or physically the width of a magnetic vortex (half a "zigzag" in longitudinally oriented media). Such zigzags in a highly oriented medium have been shown to be uncorrelated (Arnoldussen and Tong, 1986; Tang and Osse, 1987) and therefore can be considered as independent noise sources. Thus in Eq.(12) $N_v = 2/(w_c\lambda\delta)$. Since transition noise exists only in the vicinity of a magnetization reversal, "l," inside the averaging brackets, must be approximately the length of the transition. At a microscopic level we would say the noise is correlated over the length of each zigzag segment or vortex.

Guided by the observation that transition noise FNPS(k) displays an approximately exponential dependence on $|k|$, a Lorentzian function $l^2/(x^2+l^2)$ has been used to describe the real-space noise autocorrelation function (Baugh et al., 1983), "l" in this case being an average correlation length. For an arctangent transition, $M(x) = (2M_r/\pi)\arctan(x/a)$, the average magnetic charge also is distributed by a Lorentzian with shape $a^2/(x^2+a^2)$. Apart from a prefactor, the autocorrelation of this charge distribution is itself a Lorentzian with $l = 2a$, i.e., $(2a)^2/[x^2+(2a)^2]$. If the magnetization noise were distributed in proportion to the arctangent transition's average charge, then the correlation length "l" would be $2a$. Baugh et al. (1983) found experimentally $l \sim 3a$. Belk et al. (1985, 1986) found $l \sim 2a$. Since the arctangent transition shape is only an approximation, the cautious conclusion is that the noise correlation length is *proportional* to the transition length. However, Baugh et al. used a simplified formula to estimate the "a" parameters, which underestimated the values. Recomputing their "a" parameters, using the full Williams and Comstock (1972) formulation, suggests that indeed $l \sim 2a$. Moreover, Monte Carlo simulations of Lorentzian charge distributions also indicate $l \sim 2a$.

Section 3.2 of this chapter discusses intrinsic (micromagnetic) spatial noise distributions. There we see that if charge is distributed within the transition as a Gaussian, the variance of magnetization has the distribution [erf(x)][1-erf(x)], where erf(x) is the cumulative distribution of a Gaussian function. Similarly, if the charge is Lorentzian distributed the variance of magnetization is distributed as $.25\{1-[(2/\pi)\tan^{-1}(x/a)]^2\} \approx .25/[1+(x/2a)^2]$. Given this prediction, the magnetization noise (magnitude) cannot simply be distributed as the average charge, because the magnetization variance distribution would be proportional to $1/[1+(x/a)^2]^2$, which is much too narrow.

While we may not know the correct noise distribution to calculate the autocorrelation function *a priori*, we know experimentally it has a Lorentzian form (for many media), and we know the form of the variance. Inasmuch as the integral of the variance over all x should be equal to (L)[acf(0)] (see Eq.(5)), we can compute that the expected acf(x) for the magnetization noise sources is approximately $(\pi a/2)/[1+(x/2a)^2]$. Its Fourier transform is $|F(k,l)|^2$, appearing in Eq.(12), and is equal to $(\pi a)^2 \exp(-|k|2a)$. Substituting this and N_v from above into Eq.(12) we obtain the following expression for the transition flux noise power spectrum.

$$FNPS(k) \approx 2 \times \left(\frac{2}{\lambda \langle w_c \rangle \delta} \right) M_r^2 \delta^2 \langle w_c^2 \rangle \left[(\pi a)^2 e^{-|k|2a} \right]$$

$$\approx 2 D_T M_r^2 \delta \ (\pi a)^2 \frac{\langle w_c^2 \rangle}{\langle w_c \rangle} e^{-|k|2a} \tag{18a}$$

For an error function transition with Gaussian charge distribution, the analogous transition noise form is given by Eq.(18b). D_T is the linear transition density $(2/\lambda)$.

$$FNPS(k) \approx 2 \times \left(\frac{2}{\lambda \langle w_c \rangle \delta} \right) M_r^2 \delta^2 \langle w_c^2 \rangle \left[\frac{1}{4} (\pi a)^2 e^{-\frac{\pi}{4}(|k|a)^2} \right]$$

$$\approx \frac{1}{2} D_T M_r^2 \delta \ (\pi a)^2 \frac{\langle w_c^2 \rangle}{\langle w_c \rangle} e^{-\frac{\pi}{4}(|k|a)^2} \tag{18b}$$

Although Eq.(18b) assumes an error function transition shape, the "a" in both equations is the Williams and Comstock (1972) "arctangent transition parameter." The standard deviation, σ, usually appearing in an error function is set equal to $(\pi/2)^{1/2}a$. Thus the slope at the center of the transition is $2M_r/\pi a$ in both cases.

For a highly oriented film medium (Ferrier et al., 1988; Ferrier et al., 1989; Ferrier et al., 1990) noise was measured and Lorentz TEM micromagnetic images obtained. The Williams and Comstock "a" parameter for this medium was calculated to be 400 nm, in close agreement with that obtained from the images (Arnoldussen and Tong, 1986); the average w_c was found to be about 250 nm; and the measured $M_r\delta$ was 0.022 A (SI units) or 2.2 x 10^{-3} emu/cm^2 (cgs units). For these parameters Eq.(18) gives a $\delta \times$ FNPS(k$\rightarrow$0) $\approx$ (3 $\rightarrow$ 13) x 10^{-17} A^{2}m^2 for a written "all ones" pattern of fundamental wavelength 6.35 μm. (The low value is for Eq.(18b) and the high value for Eq.(18a).) The low frequency FNPS obtained from the measured spectrum as well as that computed by Lorentz image analysis were both in this range. The arctangent transition shape assumption probably overestimates the noise because real transitions do not have as long tails as the arctangent function implies.

The approximate agreement of the Eq.(18) noise with experimental measurement lends credence to the parametric dependencies implied in that equation. Since transition noise is not stationary in the statistical sense, Eq.(18) correctly shows noise power proportional to transition density (Tanaka et al., 1982; Baugh et al., 1983; Belk et al., 1986). It predicts that disk noise power will be proportional to the cross-track correlation width, w_c, of the transitions. Actually it predicts proportionality to $<w_c^2>/<w_c>$, weighting the wider zigzags more heavily, which should be expected. The noise power, like the signal power, is predicted to be proportional to $(M_r\delta)^2$. FNPS(k) should be proportional to a^2 times a function which decays with increasing k, the functional behavior being exponential, Gaussian, or something else depending on the detailed noise distribution within the transition. The total mean-squared noise, obtained by integrating FNPS(k) over all $k/2\pi$, should be proportional to transition length to the first power, as can be seen by integrating either form of Eq.(18). We also see from Eqs.(18a,b) that a transition with a narrower charge distribution (Gaussian) produces less mean-squared noise than one with a broader distribution (Lorentzian), even when the transitions have the same central slopes. In micromagnetic terms, a medium with intergranular coupling or grain orientation that favors more elongated zigzags than another may have a similar "a" parameter (central slope), and yet be noisier.

An experiment was done in which transitions were written on a disk with the recording head flying at various heights (to vary the transition length), but read always at the same flying height (Belk et al., 1986). They indeed observed that the integrated noise increased with increasing "a" parameter. However, their study was not extensive enough to establish the detailed functional relationship between noise and transition length which is predicted above. In an earlier study (Tanaka et al., 1982), pulse widths were varied by varying the write current, and the authors found that the integrated noise power varied linearly with the voltage pulse width, suggesting linearity with the transition length, but not unambiguously proving it. We must remember that, in such experiments, w_c may vary along with transition length.

122

2.3 Modulation Noise

There is frequent confusion in the literature about modulation noise, especially when discussing thin film media. For particulate media, some authors have implied that particle dispersions which do not obey Poisson statistics give rise to modulation noise. However, this is not always true because local particle correlations (non-Poisson distributed) can be regrouped as composite noise sources (Nunnelley et al., 1987), remaining additive in nature as long as their lengths are less than the average transition length or their widths are much less than a track width. For film media, inasmuch as transition noise occurs in bursts localized in the transition regions, some authors call this modulation noise because the bursts occur at the same repetition rate as the signal. This is lack of "stationarity" and is not identical to modulation noise. Similarly peak position jitter is sometimes confused with pulse position jitter, which are also not the same. The latter is a modulation noise effect (the entire pulse being shifted), whereas the former need not be (additive noise can alter the pulse locally, such as near the peak, without uniformly shifting the entire pulse). This section addresses the origins and properties of modulation noise, in contrast with additive noise.

2.3.1 Stationarity

Stationarity can have important implications for both additive and multiplicative noise, but is discussed here because of the frequent confusion between it and modulation. Recall the earlier discussion of a pedagogical noise measurement in which many (N) tracks of identical data records are written on different, but statistically similar regions of a disk. Each track is read multiple times to average out the explicit time dependent noise. If each averaged voltage record is designated $e_j(x)$, the average signal and the noise on the jth track are

$$\langle s(x) \rangle = \frac{1}{N}\sum_{j=1}^{N} e_j(x)$$
$$n_j(x) = e_j(x) - \langle s(x) \rangle \tag{19}$$

As N goes to infinity $\langle s(x)\rangle$ approaches the true signal. Many statistical functions of n(x) can be used as criteria for stationarity, three examples being the mean, mean-square, and mean autocorrelation given by Eqs.(20-22) (Bendat and Piersol, 1980).

$$\langle n(x) \rangle = \frac{1}{N} \sum_{j=1}^{N} n_j(x) \qquad (20)$$

$$\langle n^2(x) \rangle = \frac{1}{N} \sum_{j=1}^{N} n_j^2(x) \qquad (21)$$

$$R(x, x_o) = \frac{1}{N} \sum_{j=1}^{N} n_j(x) \, n_j(x+x_o) \qquad (22)$$

All such statistics must be independent of x for the noise to be stationary. The average of $R(x,x_o)$ over the domain of x is the acf(x_o) defined earlier. Equation (20) is always equal to zero, noise being random, and therefore independent of x. Equation (21), a special case of Eq.(22) when $x_o = 0$, is always a function of x for transition noise, whether additive or multiplicative. When x_o is greater than a transition length, Eq.(22) is independent of x for transition noise which is additive, but generally is dependent on x for modulation noise. While it is true that modulation noise is always non-stationary, non-stationarity does not imply that transition noise is modulation noise, despite the fact that it occurs in bursts of the same repetition rate as the signal.

As mentioned in Section 2, the concept of stationarity is mainly important for interpreting noise measurements and calculations. There is seldom any difficulty discerning the portion of a spectrum analyzer measurement which is attributable to noise, stationary or not. Similarly, Eqs.(3-5) in Section 2 are valid mathematical formulations, whether the noise is stationary or not. As stated in Section 1.1, noise is what is left when signal has been subtracted. However, if the statistical nature of the noise varies with time (non-stationarity), then it is impossible ascribe the measured spectral properties to any one noise mechanism. Noise contributions from different regions along a recorded track (with possibly different statistical characteristics) become mixed. This is the chief consequence of non-stationarity.

Yet, in practice, one mechanism is often dominant, overall or in a restricted spectral region, so that interpretation is not hopeless. For example, particulate media with mesoscopic agglomerations will exhibit noise due to the statistics of the microscopic particle fluctuations, as well as modulation noise due to large inhomogeneities. A spectral noise power measurement will mix these, but modulation sidebands can usually be distinguished if modulation is a significant contributor. For a film medium, noise may come from the transition regions as well as the regions

124

between transitions, which will have different statistical character. These become mixed in a spectral power measurement and so the spectrum cannot be attributed to "transition noise" uniquely.

In a nonrigorous sense, non-stationarity of transition noise can be thought of as a duty-cycle effect. When noise in the transitions is dominant, with the inter-transition regions being relatively quiet, the mixing of spectral contributions from noise in the transitions with inter-transition noise is inconsequential. Only the transition noise contributes, and all transitions presumably obey the same statistics. Therefore spectral interpretation is unambiguous. The measured magnitude of the transition noise requires a modified interpretation, because the noise power is averaged over both the transition length and the inter-transition length. The ratio of the transition length to the bit length is the duty cycle. The higher the bit density, the higher the duty cycle. Thus when transition noise dominates, the measured noise power is basically the average noise power within a transition times the duty-cycle factor, which is transition length times bit density. Chapter 3 cautions against blithely using a duty-cycle analogy. When significant noise occurs between transitions, such as for track-edge noise on a very narrow track, erroneous interpretation can result. However, for a theoretical treatment (as well as many real experimental conditions), we can assume inter-transition noise is zero, and the duty-cycle is a useful conceptual aid. Even then one must keep in mind that noise is not uniformly distributed within a transition, so that, as the duty-cycle becomes unity (track filled with transitions, with no inter-transition space), the transition noise remains non-stationary in the strict sense.

2.3.2 Modulation Noise Formulation

Micromagnetic fluctuations on a scale small compared to the transition length and the track width will be additive. Fluctuations of media properties (or head-media spacing, which will not be addressed here) on a scale larger than a bit cell affect the local average transition (and voltage pulse) shape. Assume a voltage pulse is a function of position, x, and various parameters such as remanent magnetization, coercivity, magnetic layer thickness, etc. Any one of these parameters, denoted as $p(x)$, has a nominal value p_o, but is assumed to fluctuate randomly with x on a sufficiently large scale. Furthermore, assume that voltage record is comprised of a superposition of basis functions $u(x,p)$ (e.g., isolated pulses), each displaced to some position x_j with alternating polarity. The signal record becomes

$$s(x) = \sum_{all\ j} (-1)^j\, u(x-x_j, p_o) \tag{23}$$

The voltage record, signal plus modulation noise, can be written as follows, using a first order Taylor expansion

$$e(x) = \sum_{all j} (-1)^{j} \left(u(x-x_j,p_o) + \left[\frac{\partial u(x-x_j,p)}{\partial p} \right]_{p_o} \Delta p(x) \right) \qquad (24)$$

The random variable $\Delta p(x)$ multiplies the derivative of the signal with respect to a macroscopic property, hence the name multiplicative noise. This derivative has the same spatial repeat pattern as the signal, therefore harmonic frequencies will be the same as for the signal. And in the frequency domain, the modulation will be seen as sidebands of noise clustered about the harmonics of the signal.

To see that this is so, it is convenient to write Eq.(23) as a Fourier series, assuming an "all ones" pattern

$$s(x,p) = \sum_{m=-\infty}^{\infty} C_m(p) \, e^{imk_o x} \qquad (25)$$

Here k_o is the signal fundamental frequency. The modulation noise can then be written

$$n_M(x) = \left[\frac{\partial s(x,p)}{\partial p} \right]_{p_o} \Delta p(x) = \sum_{m=-\infty}^{\infty} \left[\frac{\partial C_m}{\partial p} \right]_{p_o} \Delta p(x) \, e^{imk_o x} \qquad (26)$$

Upon Fourier transforming

$$N_M(k) = \sum_{m=-\infty}^{\infty} \left[\frac{\partial C_m(p)}{\partial p} \right]_{p_o} \Delta P(k-mk_o) \qquad (27)$$

And the noise power spectrum is

$$NPS(k) = \lim_{L \to \infty} \frac{1}{L} \sum_{n=-\infty}^{\infty} \sum_{m=-\infty}^{\infty} \left[\frac{\partial C_n}{\partial p} \right]_{p_o} \left[\frac{\partial C_m^*}{\partial p} \right]_{p_o} \Delta P(k-nk_o) \Delta P^*(k-mk_o) \qquad (28)$$

Suppose the variation in the modulation parameter p takes the form

$$\Delta p(x) = \sum_{q=-\infty}^{\infty} p_q \, r(x_q, l_q) \qquad (29)$$

where $r(x_q, l_q)$ is 1 for $-l_q/2 < x-x_q < l_q/2$, and zero elsewhere. l_q is the length of the modulation region of average value p_q. Substituting Eq.(29) into Eq.(28) and making use of the assumption that x_q, l_q, and p_q are random variables, we obtain

$$NPS(k) = \lim_{L \to \infty} \frac{1}{L} \sum_{n=-\infty}^{\infty} \left| \frac{\partial C_n}{\partial p} \right|^2_{P_o} \sum_{q=-\infty}^{\infty} |l_q p_q|^2 \, sinc^2[(k-nk_o) l_q/2] \qquad (30)$$

The $(k-nk_o)l_q/2$ argument of the sinc function produces sidebands about the signal harmonics nk_o. The spectral shape will depend on the distribution of modulation sites. Integrating Eq.(31) over all $k/2\pi$, the wide-band integrated noise becomes

$$\frac{1}{2\pi} \int_{-\infty}^{\infty} dk \, NPS(k) = \left\langle \left[\frac{\partial s(x,p)}{\partial p} \right]_{P_o}^2 \right\rangle_x \frac{\langle p_q^2 l_q \rangle}{\langle l_q \rangle} \qquad (31)$$

The first factor represents an average over x, and is the mean-square value of the derivative of the signal with respect to p. Since the signal consists of pulses, this is the power in a pulse (differentiated with respect to p) times the duty cycle factor discussed above, which is proportional to the transition density. The other factors represent the proper averaging over the various "q" modulation regions. Thus, modulation noise, like additive transition noise, is proportional to the transition density. However, the first factor of Eq.(31) is proportional to the signal squared, and therefore proportional to track width squared (noise $\propto$ signal adding coherently across the track width, unlike additive noise).

2.3.3 Physical Origins of Modulation Noise

As described by the foregoing equations of this section, modulation calls for a variation of some parameter, either regular or random, to multiply the time varying signal function (or its derivative). The modulating function affects the local signal function as a whole. This means that a pure modulation disturbance must be at least as wide as a trackwidth and at least as long as a transition length (or a wavelength for a sinusoidal signal). Its effect must add coherently across the trackwidth, not incoherently as for additive noise.

Historically, modulation noise has been a very real and significant effect for particulate media, especially tapes, where head-medium spacing and coating thickness variations are major factors. Sometimes particle dispersion inhomogeneities can occur on a large enough spatial scale to produce true modulation noise. Such long wavelength dispersion fluctuations produce their greatest modulation effects for low written frequencies or, in the limit, for dc-erase conditions. In the discussion of Fig.4 the spectral noise peak near k = 0 was attributed to such possible modulation. Unfortunately the terms "dc-erase noise" and "dc-modulation noise" have come to be used interchangeably. As pointed out in the discussion of Fig.4, only a portion of the dc-erase noise might be due to modulation, while the rest (the major part for well dispersed media) is additive.

Because modulation noise is sometimes associated with non-uniform particle dispersions (clumping, chaining, etc.) workers often refer to all correlation related noise as modulation noise. However, small scale correlations are more accurately treated as composite independent (additive) noise sources. Small scale variations do not multiply the average transition function, they add incoherently across the trackwidth, and they give rise to noise power proportional to trackwidth to the first power. Modulation noise sources, having a larger width and length scale, produce noise power proportional to trackwidth squared and sidebands about the signal harmonics in the frequency domain.

For film media, the term modulation noise is similarly, perhaps more frequently, misused. Most broadband transition noise in film media shows noise power proportional to track width to the first power, implying incoherent track width averaging and additive noise. This is caused by magnetic vortices and zigzag boundaries in the transition region, which are random on a small scale (micromagnetic, not meso- or macromagnetic). Of course, modulation noise can occur in film media. For film disk media, which can be nearly free of asperities that perturb the head flight, flying height variations are not a strong source of modulation noise. Nor are magnetic film thickness variations. However, the sputtering equipment and processes (e.g., substrate polishing) used to fabricate film media can frequently cause mesoscopic or macroscopic variations of local anisotropy, coercivity, or remanent or coercive squareness. These are often sufficiently long wavelength phenomena that if the primary result is amplitude modulation, the automatic gain control (AGC) of the read electronics can minimize the effect on error rate. However, some modulations are too abrupt for the AGC to correct. Moreover, due to various asymmetries (the read head, the write field near the transition center, etc.), even long wavelength modulations can become manifest as peak shifts uncorrectable by an AGC. Aside from origin and spectral distribution, the most important aspect of true modulation noise is that its mean square value is proportional to track width squared - making its impact on error rate generally more potent.

2.3.4 Modeling Insights and Pitfalls

Theoretical conclusions about noise can be subtly influenced by the modeling approach employed. Inasmuch as numerical micromagnetic models employ computationally intensive arrays, adequate modulation simulations are very difficult to do, even with the most advanced of today's computers. Therefore micromagnetic models can generally be expected to treat additive noise only. By comparison, analytical models of detailed noise mechanisms usually are formulated in terms of fluctuating mesoscopic variables (e.g., coercivity) and assumed functional dependencies of transition shapes upon these mesoscopic variables. Therefore analytical models typically handle noise as a modulation. A theory cannot simultaneously treat noise as a modulation and incorporate incoherent track width averaging to agree with experimental observation. This fundamental inconsistency flaws some otherwise rigorous analyses in the literature.

Despite such limitations, analytical modeling can generate important physical insight, especially when augmented with micromagnetic modeling. This is well exemplified in a sequence of efforts to understand the cause of supralinear noise increase with linear transition density in film media (Baugh et al., 1983). As discussed in Chapter 3, noise is linearly proportional to bit density at low densities, then rises at a supralinear rate before eventually peaking. The first attempt to explain this effect (Belk et al., 1985) employed a modulation noise model, proposing that at sufficiently high density adjacent transitions became correlated so as to avoid each other (repulsion). Studying noise spectra, Madrid and Wood (1986) pointed out that correlation effects, if any, were opposite in sign to those assumed by Belk et al., and should decrease noise unless the transitions became intrinsically noisier at high linear density. At the same time, Arnoldussen and Tong (1986) argued that magnetic interactions between adjacent transitions should produce correlations opposite to those proposed by Belk et al., and based on analysis of Lorentz micrographs of actual recorded tracks, they proposed a mechanism which would produce excess (supralinear) noise when recorded bit spacing becomes comparable to the transition lengths. Neighboring zigzag transitions of opposite polarity have a tendency to attract each other. In a fluctuating transition boundary, some microscopic regions of adjacent transitions approach each other more closely than others, and as a result of their interaction fields, these "domain boundaries" are drawn even closer than their random non-interacting spacing. This produces an amplified zigzag fluctuation and therefore additional noisiness in each transition. Arnoldussen and Tong did not quantify this proposed mechanism.

An analytical model (Barany and Bertram, 1987) was developed to quantitatively describe the onset of supralinear noise with increasing transition density, based on random coercivity fluctuations and adjacent transition coupling via demagnetization fields. The shortcomings of this model were that (a) it was in reality a modulation noise model and (b) it depended sensitively on the assumed average transition functional shape. Nevertheless this model did illustrate how fluctuations in a previously written transition could be communicated to the next transition being written, via an interaction field. Later, micromagnetic modeling (Zhu, 1991) verified that the basic mechanism, interaction field coupling of random fluctuations, does indeed account for the supralinear noise behavior (see Chapter 6). Although the Barany and Bertram model, based on modulation, was therefore flawed, its physical insight helped to interpret Zhu's numerical micromagnetic modeling.

3. Spatial Noise Distributions and Implications

As stated earlier, the common practical noise characterization method is an rms measurement which sacrifices all phase information needed to determine the spatial distribution of noise. The spatial distribution of noise, as well as the cause of that distribution, can strongly impact the effectiveness of different data detection schemes. For example, noise which exists between transitions, either on the track or at the edge of the track (Yarmchuk, 1986), will have little effect on data integrity for peak detection. For sampled voltage detection, however, such noise may be relevant, because the decision whether a zero or a one occurs in a data stream is based on several samples of the voltage near and away from the expected voltage peaks. This section is not an in-depth analysis of noise distribution effects on data detection, but rather acknowledges their importance and surveys various distributions, their physical origins, and properties.

3.1 Modulation Noise Spatial Distributions

We will look at three types of modulation noise distributions and how they are manifested when viewed as magnetization transitions, voltage pulses, and derivatives of voltage. For simplicity we will look at $M(x)$, $dM(x)/dx$, and $d^2M(x)/dx^2$ without convolving these with head response. The variance of $M(x)$ is never measured, but is of interest because theorists often calculate noise in terms of magnetization. The first derivative is of interest because it represents the noise distribution for a voltage record, which can be measured and is important in sampled voltage detection. The second derivative is important because it is the noise form which affects peak detection (Katz and Campbell, 1979).

For simplicity we will assume the magnetization transition has the form of an arctangent

$$M(x) = \frac{2M_r}{\pi} \tan^{-1}\left(\frac{x-x_o}{a}\right) \tag{32}$$

M_r is the remanent magnetization, "a" is the transition length parameter, and x_o is the center position of the transition. Figure 6 shows the shape of the magnetization transition and its first and second derivatives. We will examine three separate modulation effects, namely variations in M_r, a, and x_o. For the purpose of illustration we consider these three as independent, giving rise to simple amplitude modulation, shape or length modulation (which necessarily modulates the voltage pulse amplitude as well as shape), and simple position modulation, respectively. In reality, modulating M_r can modulate amplitude and shape; modulating coercivity can modulate shape and amplitude; combined with either of these, asymmetries or

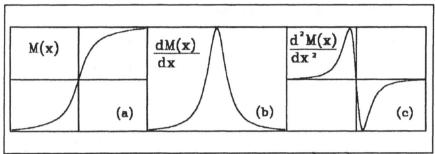

Figure 6. (a) M(x), (b) dM/dx, and (c) d^2M/dx^2 for arctan transition. Modulation noise distributions corresponding to these three views are given in Figs.7, 8, and 9.

transition interactions can modulate x_o. Such more complex effects will be combinations of the three basic effects examined here.

   Figure 7 depicts the spatial distribution of the variance (noise) associated with M(x) and its first and second derivatives due to random fluctuations of M_r on a mesoscopic scale. Similarly, Figs. 8 and 9 show the spatial distribution of modulation noise due to variations in "a" and x_o. Most importantly we note that the distribution of modulation noise depends on the form of the signal examined. For sampled voltage detection, the voltage is sampled at several points for each possible pulse to decide whether a "one" has occurred. Focusing on the (b) portions of Figs. 7 to 9, the voltage noise is distributed in proportion to the signal for simple amplitude modulation (Fig. 7) and therefore equally affects the reliability of all samples. Shape modulation puts most of the noise in the center where signal is greatest, but produces side minima and maxima (Fig.8). Simple position modulation puts the minimum noise in the center where signal is greatest and the maxima in lobes to the sides (Fig.9). Thus we expect sampled voltage detection to be sensitive to the locations of the samples, for shape and position modulation noise.

   Peak detection involves differentiating the voltage record and looking for zero crossings to locate the pulse position. For this the (c) segments of Figs.7 to 9 are relevant. Simple amplitude (M_r) modulation and shape modulation put noise peaks to the side of the signal center, while position modulation creates a major noise peak at the center. Thus it might be expected that peak detection will be most vulnerable to modulation effects that involve pulse shifting.

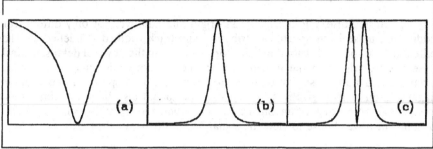

Figure 7. Noise squared spatial distributions due to amplitude (M_r) modulation for each of three views (a,b,c) of the signal in Fig.6.

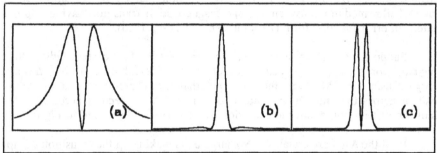

Figure 8. Noise squared spatial distributions due to transition length ("a") modulation for each of the three views of signal in Fig.6.

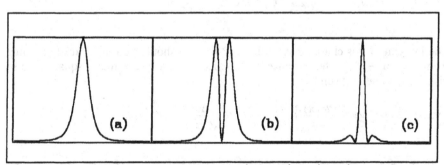

Figure 9. Noise squared spatial distributions due to transition position (x_o) modulation for each of the three view of signal in Fig.6.

3.2 Intrinsic Noise Spatial Distributions

Intrinsic noise, micromagnetic in origin and additive, not only differs from modulation noise in its spectral distribution and track width dependence, but has quite different spatial distribution behavior. Aside from the physical details, intrinsic noise has its origins in the distribution of one or more micromagnetic random variables, and how those distributions are translated into spatial distributions of magnetization or its derivatives. In essence, the spatial distribution of this type of noise remains tied to the random variable(s) producing the noise, and not to the form in which we choose to view the signal $M(x)$, dM/dx, or d^2M/dx^2.

By way of example, let us model micromagnetic noise in two simple ways, neither of which is likely to describe a real physical situation but in combination are plausible. Exploring them separately lends insight into the statistical behavior of intrinsic noise. We are guided by the observation that zigzag transitions in an oriented film medium cause an average magnetization transition having approximately an error function form (Arnoldussen and Tong, 1986).

Suppose we create a magnetic charge distribution corresponding to a magnetization transition by placing units of charge of length Δl and width Δw and charge density $q = 2M_r/\Delta l$ on the track such that each charge unit is at a different position across the track width and located in the x - direction according to a Gaussian distribution about the transition center. This is illustrated in Fig.10.

If all the Δw's are equal, the N charge units making up the transition occupy a trackwidth $W = N\Delta w$. Letting Δl be small, the average charge distribution can be written

$$\rho(x) = -\frac{dM(x)}{dx} = \frac{1}{N}\sum_{j=-\infty}^{\infty}\frac{2M_r}{\Delta l}\delta_{x,x_j} \rightarrow \frac{2M_r}{\sqrt{2\pi}\,\sigma}e^{-\frac{x^2}{2\sigma^2}} \qquad (33)$$

The integral of this charge distribution is $2M_r$, as it should be for a transition going from $-M_r$ to $+M_r$. The variance of this distribution is the mean squared noise distribution and is given by

$$var[\rho(x)] = var\left[\frac{dM(x)}{dx}\right] = \left[\frac{2M_r}{\Delta l}\right]^2 N\frac{\Delta x}{\sqrt{2\pi}\,\sigma}e^{-\frac{x^2}{2\sigma^2}} \qquad (34)$$

Δx is the finite interval over which the statistic is being sampled. For this scenario the mean square noise of the charge (voltage) distribution has the same functional form as the charge distribution itself, where charge position is the random variable. The variance of d^2M/dx^2 is

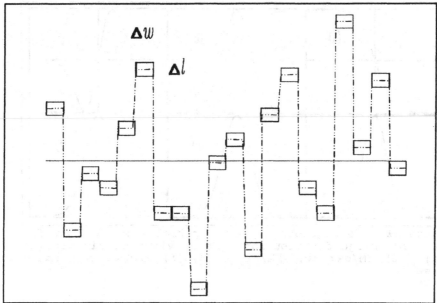

Figure 10. Transition formed by Gaussian distributed locations of magnetic charges, each of width Δw.

$$var\left[\frac{d\rho(x)}{dx}\right] = var\left[\frac{d^2M(x)}{dx^2}\right] = \left[\frac{2M_r}{\Delta l\Delta x}\right]^2 2N \frac{\Delta x}{\sqrt{2\pi}\,\sigma}\, e^{-\frac{x^2}{2\sigma^2}} \qquad (35)$$

Again the distribution of noise is the same as the original charge distribution, apart from prefactors. Despite appearances, this variance does not become infinite as $\Delta x \to 0$ because in the finite sampling $\Delta x \geq \Delta l$, and physically d^2M/dx^2 remains finite. Finally, the variance of the magnetization distribution has a slightly different functional form because the magnetization is given by the cumulative distribution of charge. It is

$$var[M(x)] = M_r^2\, N\; erf\left(\frac{x}{\sqrt{2}\,\sigma}\right)\left[1 - erf\left(\frac{x}{\sqrt{2}\,\sigma}\right)\right] \qquad (36)$$

where $erf(x/\sqrt{2}\sigma)$ is the error function, i.e., integral of a Gaussian from minus infinity to x. Although M(x) has a slightly different functional form for its variance than have the derivatives of M(x), all three forms (see Fig.11) show the noise peaking at the center of the transition. We will return to this after looking at one contrasting hypothetical distribution.

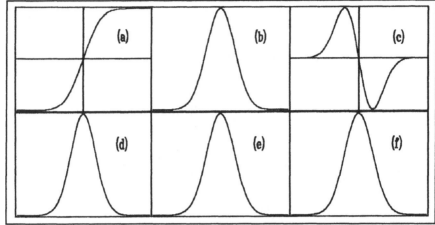

Figure 11. Mean squared noise spatial distribution for Gaussian charge distribution. Three views of signal: (a) M(x); (b) dM/dx; (c) d^2M/dx^2. (d,e,f) are corresponding noise.

Suppose we form another transition, but this time we use charge units of widths "w," lengths "l," and charge densities $2M_r/l$ where now all the charge units are centered at the transition center $x=0$, but the lengths are statistically distributed. We assume the lengths "l" have a log-normal distribution. The conclusions drawn will not essentially change for different assumptions about the "w" distribution. To maintain relative simplicity we choose "w" to be proportional to "l" (longer zigzags do tend to be wider). This is illustrated in Fig.12.

The average charge distribution is then

$$\rho(x) = \frac{2M_r}{l_o} \frac{1}{N} \sum_{j=1}^{N} r(x, l_j) \rightarrow \frac{2M_r}{l_o} \int_{\ln(2x/<l>)}^{\infty} d\ln(l/<l>) \frac{1}{\sqrt{2\pi}\sigma_l} e^{-\frac{[\ln(l/<l>)]^2}{2\sigma_l^2}}$$

$$= \frac{2M_r}{l_o}\left[1 - erf\left(\frac{\ln(2x/<l>)}{\sqrt{2}\sigma_l}\right)\right] \qquad (37)$$

where $2M_r/l_o$ is the slope of M(x) at $x=0$, N is the total number of charge units, $r(x,l_j)$ is unity for $-l_j/2 < x < l_j/2$ and zero elsewhere, $<l>$ is the average value of l, σ_l is the standard deviation for the log-normal "l" distribution. Only the positive x part of the distributions will be considered, because the negative half is the mirror image. This charge distribution is a function of the cumulative distribution of $\ln(l/<l>)$ and therefore the variance of the charge distribution is given by

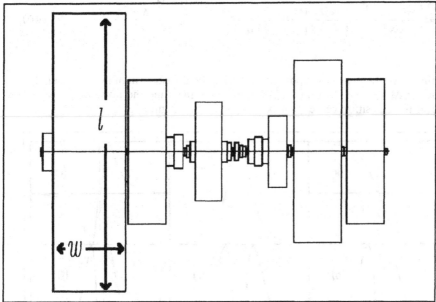

Figure 12. Transition formed by charge units all centered at x=0 but having lognormal length distribution and widths proportional to lengths.

$$var[\rho(x)] = var\left[\frac{dM(x)}{dx}\right]$$

$$= \left[\frac{2M_r}{l_o}\right]^2 N\left[1-erf\left(\frac{\ln(2x/<l>)}{\sqrt{2}\sigma_l}\right)\right]erf\left(\frac{\ln(2x/<l>)}{\sqrt{2}\sigma_l}\right) \quad (38)$$

M(x) is of course the integral of dM/dx. It can be expressed analytically but will not be written here. Suffice it to say that it is of the form $M_r c(x)$ where c(x) is a cumulative probability distribution function of the random variable having the properties of being zero at x=0, 1 as x goes to infinity, and rising rapidly near x=0. The variance of M(x) therefore peaks near x= <l> having the form

$$var[M(x)] = M_r^2 N c(x)[1 - c(x)] \quad (39)$$

The second derivative of M with respect to x (corresponding to the derivative of voltage) is proportional to the distribution of the fundamental random variable "l" and therefore the variance is

$$Var\left[\frac{d^2M(x)}{dx^2}\right] = \left[\frac{2M_r}{l_o\Delta x}\right]^2 N \frac{1}{\sqrt{2\pi}\,\sigma_l} e^{-\frac{[\ln(2x/\langle l\rangle)]^2}{2\sigma_l^2}} \frac{\Delta x}{x} \qquad (40)$$

Again, as for the previous hypothetical model, the distribution of noise is roughly the same for M(x), dM(x)/dx, and $d^2M(x)/dx^2$, only this time the noise peaks to either side of the transition center near x = <l>/2. See Figure 13.

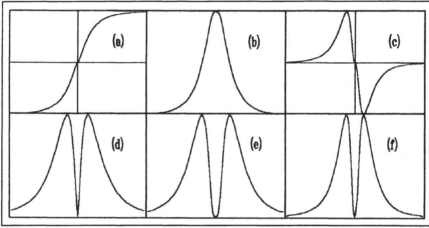

Figure 13. (d,e,f) Mean squared noise spatial distribution for centered charge blocks with lognormal length, width distributions, for three views of signal: (a) M(x); (b) dM/dx; (c) d^2M/dx^2.

Curiously, the first distribution was comprised of charge elements randomly shifted about the transition center but producing noise peaked at the center. In contrast, the second transition was comprised of charge units all centered at the transition center but with distributed lengths, and this gives rise to noise peaks near ± <l>/2. Compare these results to the modulation noise distributions seen in the previous section. There, pulse shifting produced voltage (dM/dx) noise peaks to either side of the transition center, while transition length fluctuations ("a" variations) produced voltage noise with a dominant peak at the transition center.

4. Summary

In this chapter we have reviewed definitions of various terms such as signal, distortion, interference, noise, and error rate which are frequently used imprecisely. We have derived expressions for additive and multiplicative (modulation) noise and illustrated their application to both particulate and thin film media. The physics of micromagnetics were not addressed (that is done in Chapter 6); the theoretical considerations here being limited to the global statistical aspects. An underlying focus of this chapter has been the contrast between additive noise and modulation noise. The characteristics of these two categories are summarized as follows.

4.1 Additive Noise

(1) Caused by micromagnetic fluctuations which are small compared to trackwidth and bit length;

(2) Noise power proportional to trackwidth;

(3) Spectral noise power is independent of signal (other than the duty cycle effect where integrated transition noise power is proportional to bit density) and is not manifested as sidebands of signal harmonics;

(4) Noise power has essentially the same time (or longitudinal position) distribution across a transition, regardless of whether one is viewing magnetization or its first or second derivatives;

(5) Transition noise power is proportional to linear transition density (up to the point where supralinear behavior sets in);

(6) For "zigzag" transitions having half-zigzag widths w_c statistically distributed and average transition length L_T, the media noise power should be proportional to $<w_c^2>/<w_c>$, the spectral noise power should be proportional to L_T^2, and the integrated noise power should be proportional to L_T.

4.2 Modulation Noise

(1) Caused by meso- or macromagnetic spatial variations, comparable to or larger than trackwidth and transition length;

(2) Noise power proportional to trackwidth squared;

(3) Spectral noise power is manifested as sidebands about the signal harmonics and is proportional to the signal or a derivative of the signal;

(4) Noise power has different time (or longitudinal position) distributions for $M(x)$, $dM(x)/dx$, and $d^2M(x)/dx^2$; and specific distribution depends on the mesoscopic parameter being varied;

(5) Transition noise power is proportional to linear transition density (below supralinear region).

5. Acknowledgements

I thank C.D. Mee for his support and useful suggestions during the preparation of this manuscript. I also thank my wife, Barbara, for her patience and editing assistance.

6. References

Arnoldussen, T.C. and H.C. Tong, "Zigzag Transition Profiles, Noise, and Correlation Statistics in Highly Ordered Longitudinal Film Media," *IEEE Trans. Magn.* **MAG-22**, 889 (1986).

Arratia, R.A. and H.N. Bertram, "Monte Carlo Simulation of Particulate Noise in Magnetic Recording," *IEEE Trans. Magn.* **MAG-20**, 412 (1984).

Barany A.M. and H.N. Bertram, "Transition Noise Model for Longitudinal Thin-Film Media," *IEEE Trans. Magn.* **MAG-23**, 1776 (1987).

Baugh, B.A., E.S. Murdock, and B.R. Natarajan, "Measurement of Noise in Magnetic Media," *IEEE Trans. Magn.* **MAG-19**, 1722 (1983).

Belk, N.R., P.K. George, and G.S. Mowry, "Noise in High Performance Thin-Film Longitudinal Magnetic Recording Media," *IEEE Trans. Magn.* **MAG-21**, 1350 (1985).

Belk, N.R., P.K. George, and G.S. Mowry, "Measurement of the Intrinsic Signal-to-Noise Ratio for High-performance Rigid Recording Media," *J. Appl. Phys.* **59**, 557 (1986).

Bendat, J.S., and A.G. Piersol, *Engineering Applications of Correlation and Spectral Analysis*, John Wiley and Sons, Inc., New York, Chapter 1, p.3, (1980).

Daniel, E.D., "Tape Noise in Audio Recording," *J. Audio Engin. Soc.* **20**,92 (1972).

Ferrier, R.P., F.J. Martin, T.C. Arnoldussen, and L.L. Nunnelley, "Lorentz Image-Derived Film Media Noise," *IEEE Trans. Magn.* **24**, 2709 (1988).

Ferrier, R.P., F.J. Martin, T.C. Arnoldussen, and L.L. Nunnelley, "The Determination of Transition Noise by Lorentz Electron Microscopy," *IEEE Trans. Magn.* **MAG-25**, 3387 (1989).

Ferrier, R.P., F.J. Martin, T.C. Arnoldussen, and L.L. Nunnelley, "An Examination of Transition Noise by Lorentz Electron Microscopy," *IEEE Trans. Magn.* **Mag-26**, 1536 (1990).

Katz, E.R., and T.G. Campbell, "Effect of Bitshift Distribution on Error Rate in Magnetic

Recording," *IEEE Trans. Magn.* **MAG-15**, 1050 (1979).

Madrid, M., and R. Wood, "Transition Noise in Thin Film Media," *IEEE Trans. Magn.* **MAG-22**, 892 (1986).

Mallinson, J.C., "Maximum Signal-to-Noise Ratio of a Tape Recorder," *IEEE Trans. Magn.* **MAG-5**, 182 (1969).

Mee, C.D., *The Physics of Magnetic Recording*, North-Holland Publishing Company, Amsterdam, 130-132 (1964).

Nunnelley, L.L., D.E. Heim, T.C. Arnoldussen, "Flux Noise in Particulate Media: Measurement and Interpretation," *IEEE Trans. Magn.* **MAG-23**, 1767 (1987).

Silva, T.J., and H.N. Bertram, "Magnetization Fluctuations in Uniformly Magnetized Thin-Film Recording Media," *IEEE Trans. Magn.* **MAG-26**, 3129 (1990).

Smaller, P., "Reproduce System Noise in Wide-Band Magnetic Recording System," *IEEE Trans. Magn.* **MAG-1**, 357 (1965).

Su, J.L., and M.L. Williams, "Noise in Disk Data-Recording Media," *IBM J. Res. Develp.* **18**, 570 (1974).

Tanaka, H., H. Goto, N. Shiota, and M. Yanagisawa, "Noise Characteristics in Plated Co-Ni-P Film for High Density Recording Medium," *J. Appl. Phys.* **53**, 2576 (1982).

Tang, Y.S., and L. Osse, "Zigzag Domains and Metal Film Disk Noise," *IEEE Trans. Magn.* **MAG-23**, 2371 (1987).

Tarnopolsky, G.J., H.N. Bertram, and L.T. Tran, "Magnetization Fluctuations and Characteristic Lengths for Sputtered CoP/Cr Thin-Film Media," *J. Appl. Phys.* **69**, 4730 (1991).

Thurlings, L., "Statistical Analysis of Signal and Noise in Magnetic Recording," *IEEE Trans. Magn.* **MAG-16**, 507 (1980).

Williams, M.L., and R.L. Comstock, "An Analytical Model of the Write Process in Digital Magnetic Recording," *AIP Conf. Proc.* **5**, 738 (1972).

Yarmchuk, E.J., "Spatial Structure of Media Noise in Film Disks," *IEEE Trans. Magn.* **MAG-22**, 877 (1986).

Zhu, J.G., "Noise of Interacting Transitions in Thin Film Recording Media," *IEEE Trans. Magn.* **MAG-27**, 5040 (1991).

CHAPTER 5

IMAGING METHODS FOR THE STUDY OF MICROMAGNETIC STRUCTURE

R.P.Ferrier
Department of Physics and Astronomy, The University of Glasgow
Glasgow G12 8QQ, Scotland

1. Introduction

Since the inception of data storage by magnetic recording a major feature has been the continuous drive towards the achievement of higher and higher areal densities of information storage. With current manufacturing technology the "bit area" is ~6.0μm^2 for longitudinal recording and it has been demonstrated (Jensen et al 1990, Tsang et al 1990) that improvement by a factor of 5 is already feasible technically. Thus the scale of recording has already reached the stage where an understanding of the role of the micromagnetism, and in particular the domain/domain wall structure of the medium, is central to an appreciation of the performance limitations. Of course it is not just the magnetic recording medium which is important in determining the ultimate performance of a recording system. We must also *inter alia* consider the transducers - the recording heads - which are used to write information to or read information from the recording. In this case not only do we require to characterise the static magnetisation distribution, but also the behaviour of domains/domain walls at the frequency of the read/write processes and this is already well into the MHz range.

So far, we have mentioned only the characterisation of the micromagnetism of the materials employed in magnetic recording and this is important in understanding its role in the performance which is actually achieved. However, what we must also seek to understand is how we can tailor the magnetic behaviour to improve the magnetic recording capabilities. Since the magnetic properties of a material are not in general inherent to the material itself, but rather are induced by the manner of its growth or subsequent physical processing, it is necessary for us to seek an understanding of the relationship between physical microstructure - in this context we include chemical composition on a micro scale - and magnetic microstructure. Consideration of the physical scales involved show that a spatial resolution in the range 10 - 500nm can suffice to characterise most micromagnetic structures, whereas physical microstructure requires spatial resolution in the range 0.2 - 2nm depending on the structural features of relevance. Through the Lorentz force which acts on an electron in motion through a region of magnetic induction, we have the capability in the conventional transmission electron microscope (CTEM) or its scanning equivalent (STEM) of combining the required structural studies; that this can be achieved for the same specimen area is important, particularly for fundamental studies related to magnetic recording. Thus a description of the imaging methods which have been developed for this instrumentation and a discussion of their capability for quantitative evaluation of magnetisation distributions will be a central part of this review. However, transmission electron microscopes do have their limitations and the most serious is the restriction on the specimen thickness, which typically must be less than ~100nm for ferromagnetic metals. Thus, although a thin film recording medium can be examined as an entity (provided it can be isolated from the disk without change to its properties), the physical scale of particulate and some perpendicular media and also of recording heads, renders transmission electron microscope methods largely inapplicable. Fortunately, other methods have been developed which permit micromagnetic structural characterisation of the materials we wish to examine. Among these are the methods made available by the scanning electron microscope (SEM), which was of course developed for the study of bulk specimens. Type II contrast in the backscattered signal provides information on the magnetisation within a sample to depths of the order of micrometres and Type I contrast in the secondary electron signal allows us to explore the stray magnetic field distribution beyond the surface of a ferromagnetic sample; the former is of particular relevance for studies of thin film recording heads, and the latter is applicable both to heads and media. The recent development of imaging in the SEM based on detecting the spin polarisation of the secondary electrons (the SEMPA technique) is a major advance and allows us to explore and characterise quantitatively the magnetisation distribution at the specimen surface.

Although electrons are a widely used probe of micromagnetic structure, there are other methods. Perhaps the first which is worthy of mention is the Bitter Pattern decoration technique (Bitter 1931). In its modern form using ferrofluid it is capable of high spatial resolution ~1µm. It is therefore a valuable tool and is widely used for qualitative evaluation of magnetic domain structure; it is not however capable of quantitative interpretation and since this is an aspect we will stress, the method will not be discussed further. Kerr optical microscopy is possibly the oldest method in terms of concept having been demonstrated first over a century ago. It was not widely applied, however, until the development of modern methods of digital image processing; this has revolutionised the technique and Kerr microscopy is now the most commonly available probe of micromagnetic structure. The second non-electron method we will cover is the novel technique of magnetic force microscopy (MFM) which has emerged in the last few years as an offshoot of scanning tunnelling microscopy.

In the following sections of this review we will discuss the various methods and illustrate their range of application with particular reference to the materials of relevance to magnetic recording and its continued development. There is always the tendency with microscopical methods to overstress their benefits. Here we will try to give a balanced view of advantages and disadvantages of each method and it will emerge, hopefully, that the importance is not of the techniques individually, since no one method can provide all the information which is required, but rather of the power available through their complementarity.

2. (Scanning) Transmission Electron Microscopy

The electron microscope in its modern form is perhaps the most important structural tool developed during this century, at least as far as structure determination at or near atomic resolution is concerned. Transmission and scanning transmission modes are available - frequently in the same instrument - and the diversity of interactions of an electron beam with a solid means that many different aspects of structure, including chemical composition, can be determined for the same specimen area and often simultaneously. There are many textbooks on electron microscopy and the reader is referred also to review articles by Chapman (1984), Chapman and McFadyen (1992) and Tsuno (1988), which deal selectively with topics of relevance here.

We start by examining the interaction of an electron beam with the magnetic induction of a thin ferromagnetic sample. This may be understood at two levels. Firstly, the classical approach is via the Lorentz force $\underline{F}$ on an electron travelling with velocity $\underline{v}$ in a region of magnetic induction $\underline{B}$,

$$\underline{F} = e\underline{v}\wedge\underline{B} \tag{1}$$

where e is the charge on the electron. For a film of thickness t, uniformly magnetised in the y-direction with saturation induction B_0 (Fig.1a), the Lorentz deflection angle β_L is given by

$$\beta_L = \frac{e}{h}B_o\lambda t \tag{2}$$

where λ is the electron wavelength and h is Planck's Constant. If the magnetic induction varies with both x and z in the vicinity of the wall then the local component of deflection will be given by,

$$\beta_x(x) = \frac{e\lambda}{h}\int_{-\infty}^{\infty}B_y(x,z)dz \tag{3}$$

It should be noted that $B_y(x,z)$ could arise from both magnetisation within the film and stray field beyond its surfaces; this point is of particular relevance when we discuss the

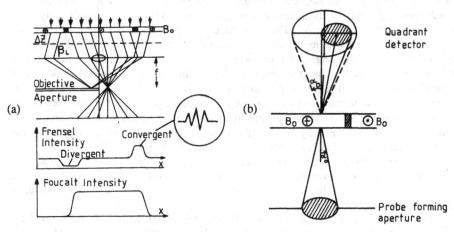

Figure 1. Schematic illustration of contrast formation in (a) the Fresnel and Foucault imaging modes and (b) the Differential Phase Contrast (DPC) mode.

imaging of written tracks in thin film recording media. Only if there is no stray field can the integration be limited to the specimen thickness. The Lorentz deflection angle β is very small; typical parameters (accelerating voltage E = 100kV; λ = 3.7pm; t = 70nm; B = 1.5T) gives a value $\beta \sim 10^{-4}$ radian which is small compared to a typical Bragg angle of $\sim 2 \times 10^{-2}$ radian. This has implications for the differentiation of magnetic and non-magnetic structure and for the high coherence of the incident electron beam which is required for Lorentz imaging.

The alternative approach to determine electron scattering by a ferromagnetic sample is using a quantum mechanical analysis and this is the preferred method when we seek to extract quantitative information on the magnetisation. Aharanov and Bohm (1959) showed that the presence of magnetic induction introduces a phase change into the electron wave. If two rays come from the same point, travel over different paths (but of the same length), enclosing a magnetic flux N and then merge again to a point, the phase shift introduced between the rays is,

$$\phi = \frac{2\pi e N}{h} \tag{4}$$

Starting at an arbitrary position x_o, the phase shift between it and a point x is given by,

$$\phi(x) = \frac{2\pi e}{h} \int_{x_0}^{x} \int_{-\infty}^{\infty} B_y(x,z) dz dx \tag{5}$$

Hence the induction distribution in a ferromagnetic film constitutes a pure phase object in the electron microscope and the imaging methods must take this into account.

We deal first with the conventional transmission electron microscope (CTEM) where, in imaging conditions, the specimen and the viewing screen are conjugate. Before describing the available imaging methods we should first note that the magnetic field inside the lens gap of an electron microscope objective lens is very high ~1T and this would provide a totally unsuitable environment for a soft or even moderately hard ferromagnetic thin film sample! There are three ways this can be overcome. Firstly, we can retain the specimen in its normal position in the microscope column and switch off the objective lens. Imaging is then performed using the next lens below in the column; this is the usual method but the image quality is limited with relatively poor spatial resolution and a magnification range which is severely restricted. A second method, but one only attainable in microscopes with a top entry stage, is to raise the specimen out of its normal position and weaken the objective excitation. Depending on the amount of movement available, the stray field at the specimen may be small enough for some studies. The third solution is the best but is expensive - it involves the use of a special objective lens polepiece (Tsuno and Inoue, 1984) designed to provide very low magnetic fields in the specimen environment (vertical fields of the order of 10^{-3}T can usually be tolerated). A schematic illustration of this polepiece, which has a double lens field, is shown later in Fig.6a.

The two modes of CTEM operation for magnetic imaging can be understood by reference to Fig.1a. Firstly, we consider the well known Fresnel mode first implemented by Hale and Fuller in 1959. Given that the induction distribution constitutes a phase object, in-focus images will have no magnetic contrast. If, however, the object plane is moved away from the specimen, i.e. we defocus the imaging lens, then phase changes are converted into intensity changes. Thus, considering the defocus plane distant at Δz below the object, the regions immediately below the domain walls will have lesser or greater number of electrons - compared to the general level - depending on the sense of Lorentz electron deflection in the specimen. If the defocus plane at Δz is imaged, the intensity will show a dip for the "divergent" wall and a peak for the "convergent" case. Since the Lorentz deflection β_L is small this means that defocus distances are relatively large ~1mm. It is also a requirement of Fresnel imaging that the coherence of the illumination is sufficiently high not to cause significant blurring. Thus the angle subtended by the effective source of illumination at the specimen must be comparable to or preferably substantially less than β_L. Given the small value of the Lorentz angle, the current density in the illuminating wave will therefore be small and it is often the case that images at only moderate magnification are invisible on the viewing screen and must be recorded "blind". This situation can be eased by incorporating an image-intensified TV camera as the recording system. This has the extra advantage that the additional magnification available extends the accessible range of magnifications with a magnetic-field-free specimen environment. In Fig.1a the quantum mechanical interpretation for the electron-specimen interaction at the convergent wall shows that it acts like a Fresnel biprism and provided the illumination coherence is sufficiently high, the wall image will comprise a set of interference fringes; in early days of Fresnel imaging these interference patterns proved valuable in assessing the validity of domain wall models by comparison of calculated and observed intensity profiles.

In all modes of Lorentz electron microscopy performed in (S)TEM instrumentation the quality of the illumination is important. In CTEM modes we have noted that contrast is adversely affected if the effective source subtends an angle appreciably larger than β_L. In scanning modes, which we will discuss later, we must ensure that the small electron probe, which ideally should be $\leq$10nm in diameter, should have sufficient current to enable images to be recorded in a reasonable time. These considerations imply that the electron gun which is used should have as high brightness as possible (brightness is defined as current cm^{-2} steradian $^{-1}$) and hence a field emission source is the preferred choice. This gun does, however, require an ultra high vacuum environment and is therefore not generally available. Hence in conventional microscopes the choice would favour an LaB$_6$ source rather than the standard thermionic tungsten cathode.

We now discuss the second method of CTEM magnetic imaging - the Foucault mode. For the situation illustrated in Fig.1a the zero order diffraction spot is split into two by the Lorentz deflections in the specimen. If an aperture in the diffraction plane is set to mask one of these spots, then only those parts of the specimen contributing to the other will be visible in the image which, unlike the Fresnel case, is in focus. The illumination coherence requirement in this case can be understood by realising that the angular separation of the magnetic spots ($2\beta_L$) must be greater than the zero order spot diameter and that is controlled by the coherence angle (strictly the angle subtended at the specimen by the effective illumination source). For a typical objective focal length of 20mm and a value β_L ~5x10^{-5} radian, the splitting will be only 1μm and hence the positioning of the aperture must be accurate to substantially better than this. For high quality Foucault images there must of course be an aperture available in a suitable plane below the specimen and this is not normally the case if the objective lens is switched off to achieve a field free specimen environment. If, however, a special polepiece is used (as discussed above) or the objective is weakly excited with the specimen raised out of the field, the required coincidence of the aperture and diffraction planes can be satisfied. It is, however, difficult to achieve quantitative microscopy with the Foucault mode.

Although we are discussing imaging modes it is convenient to break off at this point to discuss a diffraction technique which allows quantitative determination of the in-plane component of magnetic induction integrated over the electron trajectory; the technique is that of small angle electron diffraction (SAD). Again referring to Fig.1a, we see that the magnetic induction distribution in the specimen causes the zero order spot to split. Instead of arranging for the object plane to be focused, as in the imaging case, we can focus on the diffraction plane and provided we have a sufficiently long camera length we can resolve the "magnetic spots" on the viewing screen or record the pattern photographically. The requirement for the minimum suitable camera length is that the magnetic information is resolved. For a specimen with $\beta_L \sim 5 \times 10^{-5}$ radian and a spot separation at the camera of 1.0mm we need a camera length of 10.0m and this is straightforward to achieve (for a review of small angle diffraction see Ferrier (1969)). Since the camera length can be measured accurately using a suitable standard diffraction object such as a replica optical grating we can measure absolute values of β_L. This information is frequently of direct use in exploring magnetisation distributions and it is also invaluable in helping to calibrate other techniques such as the DPC method, which we shall discuss later.

So far we have discussed methods which are applicable in CTEM instrumentation. The alternative approach to imaging is the scanning transmission mode (STEM) in which a small electron probe is scanned across the specimen in raster fashion. The particular signal of interest arising from the electron-specimen interaction is detected and modulates the intensity of a CRT scanned in synchronism with the specimen scan. It has been shown that provided certain imaging parameters are matched between CTEM and STEM operation, the image information obtained from each should be the same - this is the Reciprocity Relation. Hence, in principle, there should be scanning equivalents of both the Fresnel and Foucault modes and indeed both methods can be implemented in the STEM (see Chapman and McFadyen, 1992); however they are not widely used and we will not discuss them further. The great advantage of STEM imaging is the time sequential nature of the image formation, which permits on-line image processing. For quantitative imaging of magnetisation distributions the development for STEM of the differential phase contrast (DPC) mode is probably the most significant advance in thirty years of Lorentz Microscopy. It should be noted that this method could not easily be implemented in a CTEM situation.

Examination of equation (5) shows that, since our interest is in the distribution of magnetic induction in our sample, we require the differential of the phase not the phase itself. The DPC mode was first introduced by Dekkers and de Lang (1974) for general phase objects and developed for magnetic imaging by Chapman et al (1977). The method is illustrated schematically in Fig.1b. A cone of electrons is focused to a spot on the specimen surface and scanned over it; on passing through the specimen the electron beam is deflected off axis by the local magnetic induction. The position of the bright field disc on a quadrant detector, placed on axis in the far field, will thus be displaced so that the current density distribution is not symmetrical. The amount of asymmetry can be determined by measuring the difference signals between opposite quadrants. These difference signals provide two orthogonal components of the phase differential introduced into the electron beam by the specimen, i.e. they detect the variations in the in-plane component of induction integrated along the electron trajectory. Provided that the Lorentz angle β_L is much smaller than the probe angle α, the method is to a very good approximation linear in the phase gradient, even for large values of phase change introduced by the magnetisation (Morrison and Chapman 1983). The linear approximation becomes invalid if the phase gradient is very large as would arise if we had a combination of very high saturation magnetisation and very narrow domain walls; so far this restriction has proved of little relevance to the applications of the DPC method in studying magnetic recording materials. It should also be noted that the sum of the signals falling on the individual quadrant detectors provide a standard incoherent bright field image in which physical structural information, e.g. the nature of the crystallography, the defect structure, the grain size, etc. is present in a familiar form.

To implement the DPC method there are a number of instrumental requirements. First we must form a small electron probe on the specimen, which once again should be located in magnetic-field-free space. In the VG HB5 dedicated STEM in our laboratory this is normally achieved by switching off the objective lens and using the condenser lenses for probe forming; the smallest probe which can then be formed has a diameter of ~10nm and a semiangle ~5×10^{-4} radian which, as required, is still substantially larger than a typical value of β_L. An alternative mode for higher spatial resolution is realised by moving the specimen away from its normal position and forming the electron probe by a weakly excited objective lens (in this case there will be a small d.c. magnetic field at the specimen). A probe of diameter ~3nm can be achieved with a value α~2.5×10^{-3} radian. If the image data is to be collected in a reasonable time, the probe current must be $\geq 10^{-10}$A and this can only be achieved with a field emission source. With the special objective lens (mentioned previously) in a (S)TEM system in our laboratory the probe is formed by the objective pre-field but resolution is restricted to probe sizes ~35-50nm, necessary to obtain the required probe currents with an LaB_6 source. Two other conditions must be met; the first is that descan coils are available to ensure that, in the absence of a specimen, the bright field electron cone remains centred on the quadrant detector during the image scan. Precise descanning is most difficult at low magnification and unfortunately this is often required in the study of written tracks in thin film recording media. Finally, there must be post-specimen optics to allow the bright field cone to be matched to the detector; typically the semi-angle subtended by the detector at the specimen should be ~2α. The detector itself is usually a solid state silicon p-i-n diode and quadrant forms are available commercially.

Before we proceed to discuss the applications of the DPC method, a recent improvement in experimental technique introduced by Chapman et al (1990) should be noted. If we consider DPC contrast for a real specimen non-magnetic contrast is always present and in some cases can mask the magnetic information, making it extremely difficult to extract quantitative information; this is particularly the case for the granular films currently of great interest as potential longitudinal recording media. The non-magnetic contrast arises because the specimen phase function is both magnetic and electrostatic in origin so that the differential phase for one direction should be written,

$$\nabla\phi_x = \frac{2\pi e}{h}\int_{-\infty}^{\infty}B_y dz + \frac{\pi}{\lambda e}\nabla_x(Vt) \tag{6}$$

where V and t are respectively the local electrostatic potential and film thickness. A simple but effective solution to achieving a substantial separation of the two contrast effects utilises a detector in which an annular quadrant detector surrounds a solid detector; a further sophistication would be for the central detector also to be quadrant in form. The spatial frequency content of images formed by outer and inner detectors are quite different; where low spatial frequencies are important, as is generally the case with magnetic structure, the signal to noise in the image is enhanced if the annular quadrant difference signals are used alone. The opposite is true for high spatial frequencies and since this is the case for the non magnetic features for many magnetic specimens of interest to us, a useful separation of contrast is achieved.

The applications of CTEM and STEM magnetic imaging in the investigation of magnetic recording have been mainly in the study of thin films of relevance to recording media. Investigations of physical and magnetic microstructure have been performed and the nature of the magnetisation distribution for tracks written on media has been explored. In the case of the latter it would be ideal to create the magnetic recording structure, e.g. a standard hard disk, characterise the magnetic recording behaviour including roll-off and signal to noise as a function of frequency and then isolate the thin film medium from the disk structure and perform a full physical and magnetic microstructure analysis in the electron microscope. That this has been achieved only rarely, indicates the severe problems encountered in preparing suitable microscopy specimens which retain their structure. Thus films prepared on to substrates which facilitate specimen preparation are often employed and special arrangements are necessary to record written tracks (Alexopoulos and Geiss 1985).

We illustrate the use of the electron microscope by considering a few examples of its applications in the study of thin film recording media. $Fe_{42.5}Co_{42.5}Cr_{15}$ (FCC) alloy films, prepared by evaporation at oblique incidence on to rotating NiP-coated Al-Mg hard disks (Arnoldussen et al 1984), have been investigated and their performance for longitudinal recording has been characterised. The films are magnetically anisotropic with a strongly developed easy axis of magnetisation in the tangential direction. Films isolated from the disk (Ferrier et al 1983) were first checked by energy dispersive x-ray microanalysis (EDX) to ensure that no preferential loss of elements occurred during specimen preparation. Tracks had been written on the disk with the medium in both an a.c. erased state and also saturated along the tangential direction by a large d.c. magnetic field. The Fresnel imaging mode is always the first to try since it is the most straightforward to implement. Since the domain wall contrast is insensitive to the domain magnetisation directions in the sample, the Fresnel method is the preferred mode to achieve an overall assessment of the magnetic structure. The CTEM in the Fresnel mode was used to record an image of a low frequency track written on to a d.c. erased medium, as shown in Fig.2a. As anticipated, the transition regions between bits were of the zig-zag form with an rms width ~1.3µm. What was totally unknown was the bit structure which would be

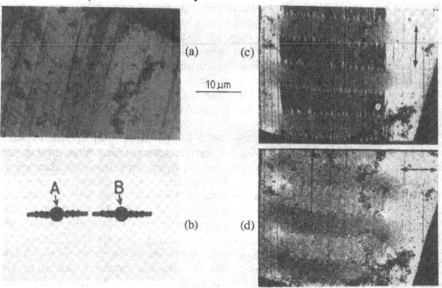

Figure 2 (a) Fresnel contrast image from a long wavelength track in a FeCoCr thin film medium (c),(d) DPC images from a similar specimen to (a) - the arrows indicate the component of magnetisation to which the image is sensitive; (b) schematic illustration of the small angle diffraction pattern which would be obtained from the specimen area as in (c). E=100keV. (Ferrier et al 1983)

formed when the frequency of recording was higher and the recording wavelength comparable to or less than the zig-zag width. The answer was that small lozenge-shaped domains were formed in the bits written in the reverse direction to the initial magnetisation as shown in Fig.3a. There is no indication in the Fresnel image of Fig.2a of magnetisation direction, although the absence of ripple contrast indicates that the magnetisation is in a constant direction. The well-defined easy axis of magnetisation in the tangential direction would imply that the magnetisation should be parallel or antiparallel to the track direction. This can be confirmed in two ways, the first being by small angle electron diffraction. The pattern which would be obtained from a region which contains portions of both track and dc erased specimen, is illustrated schematically in Fig.2b. This shows that the zero order beam is split into two strong spots and through each and collinear are two fainter diffraction streaks. Spot A can be determined as coming from the dc-erased region and spot B from the track; it may be shown in a separate experiment that spot B coincides fairly closely with the zero order diffraction spot, which would be formed in the absence of a specimen. The streaking comes from small angle scattering from thickness variations in the specimen, which arise from replication of the tangential polishing marks on the substrate surface; these are also clearly visible in the Fresnel image. The collinearity of the streaks and the strong diffraction spots indicates the magnetisation is constrained to lie in the tangential direction also. The coincidence of spot B and the zero order position shows that to first order the phase change arising from the magnetisation within a written bit is almost exactly cancelled over much of the track by the induction arising from the stray magnetic field above and below the specimen. It should be noted that if the camera length is determined, then the position of A relative to the origin will give a measure of $B_o t$. An alternative and more satisfactory method is to examine the small angle diffraction pattern from an ac-erased region of the specimen. This exhibits long narrow domains of opposite magnetisation and hence the zero order will be split into two strong spots the separation of which gives a value for $2B_o t$. (Ferrier et al 1987)

. DPC images are also very instructive and a pair of images from the low frequency track are shown in Fig.2c,d; the directions of magnetisation to which the contrast in each is sensitive is indicated by the arrows (this will be normal to the direction of differentiation). Grey contrast - 127 on an 8 bit intensity scale - indicates that there is no net component of induction normal to the electron trajectory. Thus the d.c. erased part of the specimen shows strong contrast in Fig.2c, but is grey in Fig.2d as we would expect from arguments given above. The very low contrast difference between neighbouring bits along the centre of the track in Fig.2c confirms the almost complete cancellation of the electron phase due to magnetisation inside the written bit by the strong magnetic field emanating from the zig-zag boundaries. Thus the film as a recording medium is very efficient and little is lost by flux closure at the zig-zag structure. Away from the centre line of the track, the contrast changes which are observed indicate changes in the stray magnetic field, which falls slowly in magnitude towards the edge of the track and then rapidly beyond the edge; this behaviour is expected and in agreement with results from modelling of the track magnetisation (McVitie and Ferrier, 1991). The concentration of contrast in Fig.2d only in the vicinity of the corners of the written bits arises also from the stray field configuration; the image also indicates that the magnetisation in the film is almost entirely in the track direction. The DPC image corresponding to the Fresnel image of the high frequency track of Fig.3a is shown in Fig.3b, the contrast being sensitive to induction in the track direction. In this image it is instructive to note that the contrast in the bits which have magnetisation in the same sense as the dc-erased region indicates that the stray field is quite smoothly varying despite the highly irregular structure of the other bits; this would imply that the recording noise might not be as bad as would at first be suspected from the nature of the structure and qualitatively this is in agreement with noise measurements.

That quantitative information on induction distribution may be derived from DPC images is not in doubt. However, if we are concerned to determine the magnetisation distribution for the track and this is the quantity which we require if we are to calculate the contribution of the medium to recording noise (Ferrier et al, 1987, 1988, 1990a; Arnoldussen 1992), then we must determine either the stray magnetic field distribution separately and subtract it out, or find another way of removing its influence. Certainly this is the greatest challenge to be addressed in the drive towards quantitative Lorentz Microscopy of recording media.

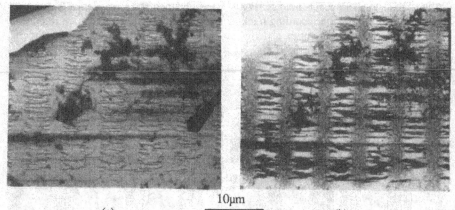

(a)   10μm   (b)

Figure 3 (a) Fresnel contrast image for a short wavelength track in a FeCoCr thin film medium, (b) DPC image for a similar specimen with contrast sensitive to the component of magnetic induction in the track direction. E=100keV. (Ferrier et al 1983)

We have mentioned that physical microstructure is important in determining the magnetic properties of materials. An example of this is the role of physical and chemical microstructure in determining the perpendicular anisotropy for CoCr films required to sustain perpendicular recording. Groups at Glasgow University and Philips Laboratories (Chapman et al, 1986; Rogers et al, 1989) have shown that sputtered $Co_{78.5}Cr_{21.5}$ grown on polyester films exhibits columnar grain structure with the c-axis of the hexagonal crystallites directed along the column axis. For perpendicular recording the alloy film thickness was 0.4μm, much too thick for examination at conventional accelerating voltages (100-200kV) in the electron microscope. Thus cross-section and planar specimens were prepared and images from the former clearly showed the columnar structure. In the vicinity of the particular CoCr composition used, the magnetic properties are a strong function of Cr content and CoCr becomes paramagnetic at a composition $Co_{75}Cr_{25}$. There was therefore the question of preferential segregation of Cr to the column boundaries, since this would enhance decoupling of the exchange interaction between the columns. To achieve meaningful results, high spatial resolution x-ray microanalytical analyses were required; the VG HB5 STEM with its field emission source provided sufficient current into small (≤2nm diameter) electron probes to carry out the experiment and the results clearly demonstrated an enrichment of ~5% at the column boundaries.

It was mentioned earlier that in principle it is possible to extract quantitative information on magnetic structure using Foucault microscopy. In practice, however, the greatest value of this imaging method is as an adjunct to Fresnel imaging since it permits relatively rapid qualitative assessment of the nature of the magnetic induction distribution in the specimen (as opposed to domain wall delineation). To illustrate this we show an example from the work of Alexopoulos and Geiss (1985) who investigated the physical and magnetic structure of sputtered cobalt alloys and their use as longitudinal recording media. Fresnel and Foucault images (Fig.4) were obtained for a magnetically isotropic film of $Co_{80}Cr_{20}$ alloy with recording tracks written on to it. The bit transition regions in Fresnel contrast show both vortex and zig-zag character but are quite difficult to discern; in the Foucault mode the transition regions are much more clearly delineated.

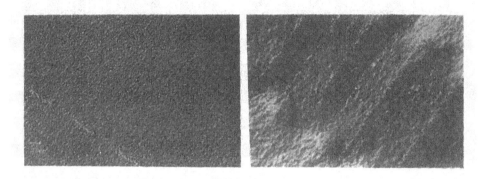

(a)   (b)

Figure 4 Lorentz micrographs of a track in a thin $Co_{80}Cr_{20}$ alloy film, (a) Fresnel contrast, (b) Foucault contrast. E=100keV. (Alexopoulos and Geiss 1985)

The potential of annular quadrant detectors to enhance magnetic contrast in DPC images at the expense of non-magnetic contrast, has been the subject of theoretical and experimental investigation (Chapman et al 1990; Chapman and McFadyen 1992). The benefit to be gained will depend on the nature of the specimen structure; we illustrate its use in the study of a 60nm thick permalloy film prepared by evaporation. For this film the crystallites exhibit strong contrast in the solid quadrant detector image shown in Fig.5a and this results in poor definition of the domain wall boundary. The improvement obtained using the annular quadrant detector (Fig.5b) is striking and the clearly defined domain wall boundary now permits a realistic estimation of the domain wall width.

In this section we have discussed the methods of transmission and scanning transmission electron microscopy which permit the investigation of micromagnetism of thin film structures and correlation with microstructure. One notable omission however is the method of electron beam holography. The use of this method to study magnetic materials has been demonstrated (for a review see Tonomura 1988). However, although the technique is exceptionally elegant, the stringent instrumental requirements are likely to mean that it will not be practised widely, hence we omit detailed discussion.

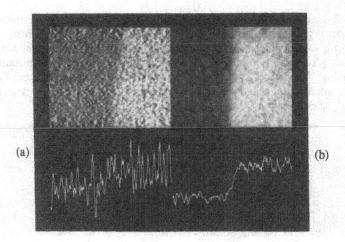

Figure 5 Comparison of DPC images and intensity line traces obtained using (a) a solid quadrant detector, (b) an annular quadrant detector. The specimen was a 60nm permalloy film. E=100keV. (Chapman and McFadyen 1992)

3. Scanning Electron Microscopy

3.1 Type II Imaging

In the SEM, Type II magnetic contrast in the backscattered electron image from bulk ferromagnetic specimens arises from changes in the effective backscattering coefficient due to the interaction of the scattered electrons within the sample and the local magnetic induction - for a recent review of Type II imaging see Tsuno (1988). Depending on the tilt of the specimen and the disposition of the backscattered electron detector(s), contrast due to the presence of either or both domains and domain walls can be obtained. In the case of the latter, the standard geometry is for the specimen to be normal to the incident beam and the detectors are positioned above it and close to the optic axis; this situation is illustrated schematically in Fig.6a. This geometry can be achieved in both SEM and STEM instruments and, as we shall see later, the greater accelerating voltage range available in the latter is a distinct advantage for the study of the micromagnetism of thin film heads, the main area of application in magnetic recording.

The main problem with imaging using Type II contrast is its inherent weakness. For example, for permalloy, the domain wall contrast at 30kV is only ~0.05% (Ikuta and Shimizu 1983), although there is a fairly rapid increase with accelerating voltage E as $E^{1.5}$. This means that although signal processing can be used to suppress the large background intensity, incident probe currents must be large if recording times are not to be too excessive. At 100kV accelerating voltage a typical probe current would be ~10^{-8}A and with an LaB_6 source and reasonable probe forming optics this would give a probe size of ~100nm; given the nature of the electron scattering in the specimen however, the observed resolution would be in the range 0.5 - 1.0μm. The major problem with Type II imaging is the influence of specimen topography - unless the specimen is smooth and flat, topographic contrast will dominate the backscattered image. Type II images have been observed for the P1 polepieces of thin film recording heads (Mee 1980), the specimen surface being sufficiently flat by virtue of the quality of the ceramic substrate. However, to the author's knowledge, no Type II images have been observed directly for the P2 polepieces of thin film heads; the reason is that although the surface is relatively smooth it is certainly not flat.

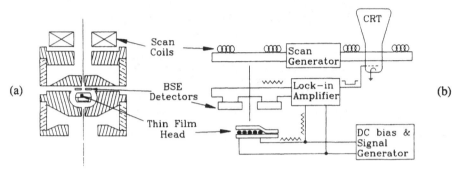

Figure 6 (a) Schematic illustration of the specimen region in the JEOL 2000FX fitted with the special double gap objective lens. (b) The experimental arrangement for the synchronous detection of the backscattered signal. (Ferrier et al 1990, c IEEE)

A major advance in experimental method for the investigation of thin film heads by Type II contrast and one which opened up the possibility of the study of domain wall dynamics was introduced by Wells and Savoy in 1981. The essence of the method was to drive the head coil with an a.c. voltage and detect that part of the backscattered signal which was synchronous using a lock-in amplifier. This does not increase the inherent contrast due to the magnetic interaction but rather suppresses to a large extent contrast arising from topographic features. Analysis of the method shows that if the specimen/detector geometry is set for domain wall observation, then in the lock-in image the extent of domain wall motion is delineated by a pair of black and white lines; only if the domain wall movement is larger than the spatial resolution can the presence of a wall be detected. In the original implementation Wells and Savoy used a standard SEM operating at a maximum accelerating voltage of 30kV and this meant that their observations were restricted to heads where the production was stopped after the formation of the P2 layer and before the deposition of the 20μm alumina overlayer. As noted above, the geometry for Type II detection can be implemented in a (S)TEM machine and this was carried out by Ferrier and Geiss (1986), who modified a Philips 301 instrument. The magnetic-field-free environment required for the soft magnetic specimens was achieved by switching off the objective lens and forming the electron probe with the second condenser lens; electron optically this arrangement is poor and hence spatial resolution was limited. However, the advantage of the higher accelerating voltage in providing a degree of depth profiling of the magnetic structure in the specimen was demonstrated clearly in studies of a P2 polepiece which exhibited two quite different magnetic structures in the two plated layers from which it was constructed. The differentiation was achieved by a combination of accelerating voltage and head drive current variation as shown in Fig.7. At low accelerating voltage and low drive current the walls which are easily visible are in the top layer of P2 and show that the anisotropy in this layer is in the "wrong" direction. At high drive currents the top layer magnetisation saturates and at high accelerating voltages the less mobile domain structure in the lower layer is clearly visible.

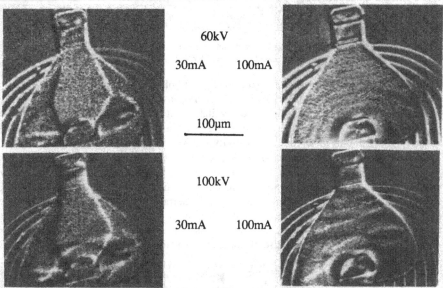

Figure 7 Determination of the different domain structures in the 2 plated layers of a P2 polepiece. The head drive currents and the accelerating voltages are shown. The drive frequency was 100kHz.

The advantages of high accelerating voltage and good electron probe optics have been combined in one instrument (Ferrier et al 1990b). The basic microscope was a 200kV (S)TEM (JEOL 2000FX) with a special objective polepiece in which the lens field was split using separate gaps and a region in the middle provided a magnetic-field-free environment for the specimen. This is illustrated in Fig.6, which also shows the arrangement for lock-in imaging. To date, the frequency of operation is limited to 100kHz by the lock-in amplifier; extension into the MHz range is highly desirable but would require high frequency lock-in amplifiers with adequate output bandwidth and an improved solid state detector/preamplifier combination. Wells and Savoy 1981) showed that for tilted specimens domain contrast effects are also visible in the synchronous detection mode and this can be exploited (Ferrier et al 1990b) to gain information on the changes in magnetisation direction over the period of the head drive current as illustrated by comparing the images of a tilted and untilted specimen in Fig.8. These show that

156

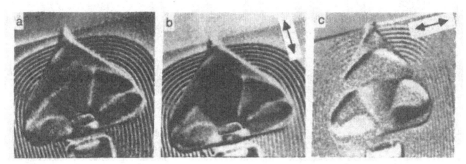

Figure 8 Synchronous detection backscattered images for a P2 polepiece (a) untilted specimen; (b),(c) tilted through 30° about the indicated axis. E=100keV, f=100kHz. (Ferrier et al 1990, c IEEE)

additional contrast, which is dependent on the axis of tilt, is visible particularly in the vicinity of the closure and reflects the large swing in magnetisation direction over this region. No quantitative analysis is yet available but the result serves to confirm the observations of large magnetisation rotation observed in lock-in Kerr imaging (see Fig.22). A striking illustration of the depth sensitivity of Type II lock-in imaging is provided by studies of planar thin film recording heads (Ferrier 1991). In this head structure the two polepieces are in the surface plane and a third lower polepiece at a depth ~13 - 15µm provides a flux closure path. Type II studies as a function of accelerating voltage (Fig.9) show that at 60 - 80kV the domain structure in the surface polepieces is detected whereas at 200kV the contrast is dominated by the domain structure of the lower polepiece; at intermediate accelerating voltages both sets of domain structures are visible simultaneously. The ability to discern magnetic structures deep below the specimen

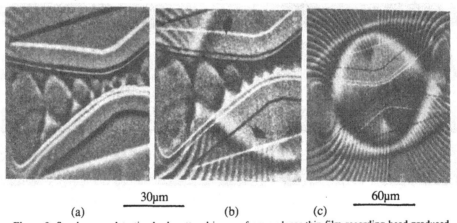

<table>
<tr><td>30µm</td><td></td><td>60µm</td></tr>
<tr><td>(a)</td><td>(b)</td><td>(c)</td></tr>
</table>

Figure 9 Synchronous detection backscattered images from a planar thin film recording head produced by LETI, Grenoble. (a) E=60keV, drive current=10mA p - p, (b) E=100keV, drive current=10mA p - p; note the appearance of domain walls in the lower polepiece as indicated by arrows, (c) E=200keV. drive current=20mA p - p. (Ferrier 1991)

surface means that Type II images can be obtained from heads with the 20µm alumina overlayer in place and this permits direct comparison with results of Kerr microscopy.

We conclude this section by discussing a new imaging method which permits the observation of Type II magnetic contrast in d.c. conditions, even in the presence of relatively large topographic contrast. Using digital image recording at 16 bit accuracy, Ferrier and McVitie (1991) have shown that if two images of a P2 polepiece are recorded under different conditions of d.c. drive e.g. one image of the remanent state and the other with a d.c. voltage applied to the head coil, then subtraction of the two images followed by contrast expansion reveals the original position of domain walls and the positions to which they have moved provided the extent of movement is greater then the spatial resolution for magnetic contrast. This is illustrated in Fig.10 and it can be seen that pairs of wall positions are in opposite contrast. This new method is not restricted to recording heads and the two d.c. states can be achieved using an external magnetic field applied in situ. The reason the method is successful is that topographic contrast will not be influenced by the magnetic field and hence cancels out in the subtracted image; the method is in fact the d.c. equivalent of the lock-in technique.

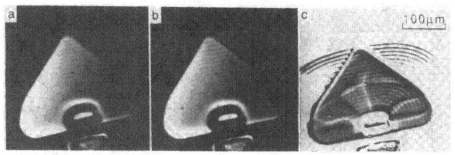

Figure 10 Backscattered images of a P2 polepiece, (a) d.c. drive current 0.0mA, (b) d.c. drive current 30mA, (c) image derived from image subtraction of (a) and (b). E=100keV. (Ferrier and McVitie 1991)

Type II imaging using both the synchronous detection mode and the image subtraction method provides valuable information on magnetic structure and it is of particular relevance to the study of thin film recording heads. The information gained is similar in character to that explored using Kerr microscopy methods and, given that the latter is more straightforward to apply, it might be regarded as the preferred technique. However, it must be recalled that the Kerr effect is essentially a probe of surface magnetisation and hence the depth profiling capability for magnetic structural studies of the Type II method provides a unique advantage. It is the author's view that it is the complementary nature of Type II and Kerr imaging which in the future will prove the greatest advantage.

3.2 Type I Imaging

In the SEM, the trajectories of secondary electrons emitted from the surface of a ferromagnetic sample will be influenced by the presence of stray magnetic fields above the surface. This is illustrated in Fig.11a for the case of a written track in a longitudinal recording medium. The asymmetry introduced into the secondary distribution will be a function of the probe position on the sample and if we employ a directionally sensitive detector then magnetic contrast known as Type I can result. Type I contrast is generally weak and special care is needed to detect its presence with any reasonable efficiency. A number of experimental arrangements have been explored and a brief review will be found in Ferrier and Kyser (1973). Here we describe one approach based on original work by Nixon and Banbury (1969) and later supplemented by theoretical analysis conducted by Munro (1971). The principle of the detector is based on ensuring the secondary trajectories are planar in the absence of any stray magnetic field. This is achieved (Fig.11b) by employing a cylindrical detector to define an electrostatic field distribution of cylindrical symmetry in the vicinity of the specimen. The specimen, which is untilted, sits immediately below the bottom plate of the cylinder. A proportion of the secondary electron distribution passes through a mesh sector in the cylinder and is detected by a scintillator photomultiplier assembly; for a suitable orientation of the specimen relative to the detector, magnetic contrast is obtained.

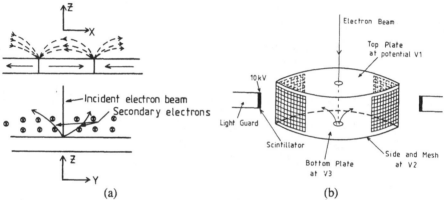

(a) (b)

Figure 11 (a) Illustration of the influence of stray magnetic field on the secondary electron trajectories. (b) Schematic of the modified Nixon-Banbury detector, $V_2 = 50V$, $V_1 = V_3 = 0.0V$ (Ferrier et al 1986)

The potentials on the cylindrical detector are tuned to optimise the signal to noise ratio in the image by selecting the portion of the secondary electron distribution which can reach the scintillator (Ferrier et el 1986). To reduce recording times to acceptable levels a high electron probe current is necessary and this means the spatial resolution will be limited; the precise value will depend on the nature of the electron gun and the performance of the probe-forming lens. In a Hitachi S800 field emission SEM a resolution ~100nm was achieved. There is a distinct advantage in having two, rather than one, mesh detectors positioned at 180° to each other. Subtraction of the signals from the two photomultipliers gives a doubling of the magnetic contrast and that due to atomic number is suppressed. Unfortunately topographic contrast is also enhanced by the method and so specimen surfaces must be as smooth as possible. The deflection experienced by a secondary electron will be given by the integral along the trajectory of the component of magnetic flux normal to the electron path. This flux-times-distance integral will be related to the $M_r\delta$ product of the medium and the larger this product the more readily visible will be the magnetic contrast. Examples of Type I contrast from thin film magnetic recording media are shown in Fig.12. In this investigation the shortest wavelength recorded track which was resolved had 2,000 flux reversals per millimetre. For this case, given that the playback signal would be well down the roll-off curve, the "effective" value of $M_r\delta$ was almost certainly substantially less than the maximum possible value .

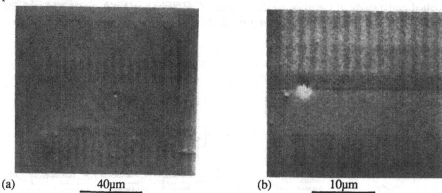

(a)   40µm   (b)   10µm

Figure 12 Type I contrast images from tracks written in thin film recording media (a) CoCr film $M_r\delta=0.003$ emu cm^{-2}, (b) sputtered Fe_2O_3 film $M_r=0.0016$ emu cm^{-2}. (Ferrier et al 1986)

If it was possible to extract quantitative information on the three dimensional nature of the stray field above a recorded track from Type I imaging, then the importance of the method in the investigation of the micromagnetism of recording would be greatly enhanced. Wells (1985) has carried out a theoretical analysis of the situation and his results indicate there is a solution. If the secondary electron distribution is measured by four detectors disposed around the optic axis then by combining the individual signals and differentiating the resultants, it is possible in theory to determine the distribution of the outward normal component of magnetic induction at the specimen surface. Given that there are no free magnetic poles in the space above the specimen this data is in principle sufficient to determine the three dimensional nature of the magnetic induction in the space beyond the specimen surface. Regretfully, no experimental implementation of the Wells proposals has been reported and Type I contrast remains qualitative and of limited utility in micromagnetic studies.

3.3 Electron Beam Tomography

Type I contrast discussed above is not the only way in which stray magnetic fields beyond the surface of a ferromagnetic specimen can be investigated with SEM/STEM

160

instrumentation. In fact there have been over the years a number of investigations all based on detecting the deviation of electron trajectories on interaction with the magnetic field - for a brief review the reader is referred to a paper by Wells (1985). Unfortunately, none of the methods provided quantitative data. Over the last few years a group at the University of Duisburg (Elsbrock et al 1985: Steck et al 1990), have pioneered a new approach and have been successful in achieving quantitative characterisation of stray fields. Whilst the basic experimental method is similar to the others and relies on measuring electron beam deflections, the data set explores the whole of the field space and is the input for a reconstruction method based on tomography. The method has been applied in the study of the microfields from thin film recording heads and a schematic of the experimental arrangement is shown in Fig.13. The recording head is placed in an

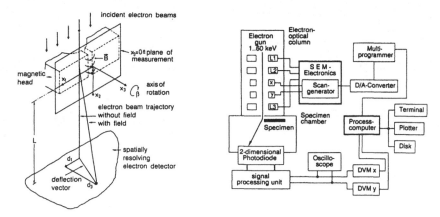

Figure 13 Schematic of the magnetic microfield determination method. (Elsbrock et al 1985, c IEEE)

SEM and the electron beam is scanned over the region of space beyond the poletip surface. On penetrating the magnetic field of the head, produced by d.c. excitation of the head coil, the electron beam is deviated and the magnitude of the deviation can be determined from the beam position on a two dimensional photodiode detector placed in a plane distant from the specimen (the beam position on the detector for the zero field case acts as a reference). The deflection vector for a two dimensional scan over the space outside the head could be recorded but it is in fact sufficient to restrict data collection to a line scan close to the poletip surface. A single scan for this plane provides information on the line integral of the component of induction normal to the trajectory. To derive the vector nature of the induction over the measurement plane it is first necessary to obtain a number of different "views" of the induction distribution in that plane. This is achieved by rotating the specimen about the normal to the poletip surface and recording line scans at angular intervals over a total rotation of 180°. This set of line scan data is input to a tomographic reconstruction procedure which gives the value of $\underline{B}$ over the measurement plane. Since there are no free magnetic poles in the space $x_3 > 0$, the calculated values of $\underline{B}$ in the plane can be used to determine the values of $\underline{B}$ in any other plane and suitable algorithms have been developed for the purpose. Simulation calculations indicate that a total of 45 projection angles with 61 scan positions along each line will give 5% accuracy in the determination of $\underline{B}$. An example of the reconstructed magnetic field near the poleface of thin film recording head is shown in Fig.14. The spatial resolution and sensitivity of the method are better than 0.1μm and 1mT respectively. The major experimental difficulty lies in the construction of the specimen stage (Steck 1990), which must permit highly accurate alignment of the poleface surface parallel to the optic axis and of rotation of the head about the required axis for the varying projection conditions.

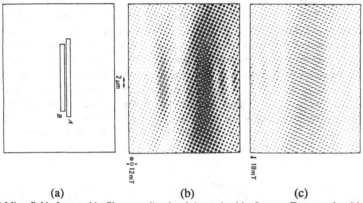

(a)   (b)   (c)

Figure 14 Microfields from a thin film recording head determined by Lorentz Tomography, (a) location of the poletips, (b) in-plane field $B_{1,2}$ (c) perpendicular field B_3. The field values are determined for a plane distant 3.0μm from air bearing surface. (Steck et al 1990, c IEEE)

An alternative approach has been adopted by Matsuda et al (1990). Their experimental arrangement is essentially the same as above, but two dimensional scans are made at each projection angle and the deflection vector due to the head field is measured as a function of both x_1 and x_3. The reconstruction procedure uses an interactive approach based on the algebraic reconstruction technique (ART) (Gordon 1974). Although 2-D data is collected, the analysis is carried out by considering planes parallel to the poleface as separate reconstruction planes. The essence of the reconstruction procedure is as follows. For the first projection, each point $(x_1,x_2,x_3 = const)$in the reconstruction plane a value for the induction component is assigned based on an average value assessed from the measured deflection vector. For the next projection, the deflection at a point on the line scan can be computed using the initial field values and compared with the measured value; the difference between the two is then used to refine the first estimate of the field values and so on. The authors claim that their iterative procedure overcomes problems which could arise with the Duisburg method due to the use of linear approximation theory.

Very few analyses have so far been carried out using these tomographic methods, but their importance must increase as recording heads get smaller in an effort to increase areal

162

recording density. In addition, there may well be scope for applying tomographic methods in the other areas of micromagnetic characterisation.

3.4 Scanning Electron Microscopy With Polarisation Analysis

As we have already discussed, the scanning electron microscope allows two "conventional" methods of studying magnetic microstructure, either through the interaction of the secondary electrons with the stray magnetic field above the specimen surface (Type I) or, in the case of backscattered electrons, via the change in effective backscattering coefficient arising from interaction of the scattered electrons with the local magnetisation in the sample (Type II). In 1976 Chrobrock and Hofmann found that the secondary electrons emitted from the surface of a ferromagnet are spin polarised, i.e. they have a preferred direction of their spin which depends on the local direction of magnetisation at the point of their creation at the sample surface (see Fig.15). The effectiveness of the process depends on the energy of the emitted secondaries and is a maximum for low energies. The acronym SEMPA (scanning electron microscopy with polarisation analysis) has been adopted for the method and details of the physics of the method and its implementation and exploitation are to be found in the recent review articles of Scheinfein et al (1990) and by Oepen and Kirschner (1991). In the following paragraphs, we discuss the main aspects of the method.

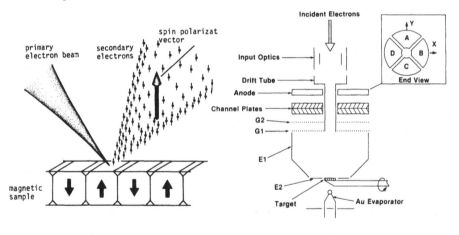

Fig.15   Fig.16

Figure 15 Principle of magnetic structure analysis by SEMPA. The spin polarisation in the emitted secondaries is dependent on the specimen magnetisation. (Oepen and Kirschner 1991)

Figure 16 A cross section of LEDS Mott detector. The divided anode assembly is shown in inset as viewed from the Au target. (Unguris et al 1986)

The most important feature of the design of the SEMPA system is the method of detecting the spin polarisation of the secondaries; this vector quantity is normally determined by measuring orthogonal components, one in the direction of propagation and the other two in the plane normal to this. The detector used must be sensitive to electron spin, and for most detectors this is achieved using the spin-orbit interaction; when an electron is scattered from the central potential of a high atomic number element, there is an additional term in the Hamiltonian resulting from the interaction of the electron spin with its orbital angular momentum in the central field. This is known as Mott scattering and the cross section is larger or smaller for electrons with spin parallel or antiparallel to the normal to the detector scattering plane. An example of one such Mott detector, which has a relatively high efficiency compared to others (the absolute efficiency is however low) is the low energy diffuse scattering (LEDS) detector illustrated schematically in Fig.16 (Unguris et al 1986). This operates at an energy of 150eV and uses an evaporated polycrystalline gold film as the target. The potentials on the electrodes E_1, E_2 are selected to make the trajectories of the electrons scattered from the target normal to the control grids G_1, G_2, which discriminate against secondaries created in the gold target. After G_2, each electron is multiplied by the channel plate and detected by a high voltage anode in the form of a quadrant. Combining the signals from the quadrants gives the two components of polarisation

$$P_x \propto \frac{N_C - N_A}{N_C + N_A} \qquad P_y \propto \frac{N_B - N_D}{N_B + N_D} \qquad (7)$$

The third component of the polarisation is determined using a second detector channel at right angles to the first.

Since this method relies on the spin polarisation of the secondary electrons at the point of their creation in the ferromagnetic sample, the depth sensitivity of the method will be determined by the escape depth of secondary electrons. This parameter has been studied by Seiler (1983) who found the escape depth to vary in the range 0.5 - 1.5nm for metals and this range has been confirmed in spin polarised studies (Abraham and Hopster 1987). Thus SEMPA is essentially a surface technique probing magnetisation distributions at or very near the surface; this therefore determines both the strengths and the limitations of the method. The surface nature of SEMPA also implies that any contamination of the virgin surface will lead to a rapid loss of information on the nature of the magnetisation; for example, Allenspach et al (1985) have shown that for iron, one monolayer of oxygen reduces the polarisation by a factor 2. Thus a UHV specimen environment is essential to the method and the normal surface physics cleaning tools such as ion beam etching should be available in situ.

The practical spatial resolution which can be achieved in SEMPA is determined by a number of factors. The most important is the required degree of sensitivity to changes in the polarisation direction, since this determines the number of detected electrons per pixel required to give a satisfactory signal to noise ratio in the image contrast. In turn this, together with the limited efficiency of the detector and the magnitude of magnetisation of the sample, determines the current required in the electron probe incident on the specimen, given that there is a practical limit to the time over which the image can be acquired due to instrumental instabilities. Whilst very small electron probes can be achieved in SEMPA instruments, the crucial parameter is the current in the probe; the probe size appropriate to a given current will be determined by the optical characteristics of the probe-forming lens and the nature of the electron gun. In general, for probes less than 100nm in diameter, an FEG is required whereas a thermionic LaB_6 source would be preferred for larger probes. In SEMPA machines the probe forming lens typically has a working distance ~10mm and is non-immersion in character; it is thus not of the highest performance. Taking all these factors into account gives a practical resolution limit ~30 - 40nm, although many studies can be carried out usefully with much larger probes. Finally it should be noted that, as in other methods discussed, modern image processing techniques play an important role in optimising the determination of sample magnetisation information. An example of this is the degradation of image quality introduced by scan-generated instrumental asymmetries and, more importantly, by aberrant electron trajectories resulting from large surface topography effects. These problems can be reduced significantly by including an additional graphite detector target which produces images equivalent to those of the gold target, but with the contrast from the magnetisation largely absent. Subtraction of these graphite images from the corresponding ones from the gold target leads to significant improvement in the visibility of the magnetic structure (see Fig.19).

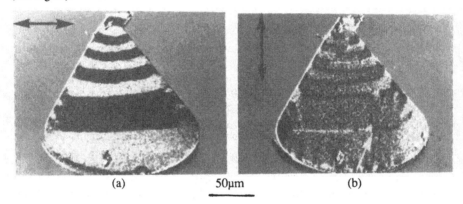

(a)   50µm   (b)

Figure 17a,b SEMPA images from a P2 polepiece. the components of magnetisation which are sensed are indicated by the arrows. Note in (b) the sensitivity to the chirality of the wall at the specimen surface. (Scheinfein et al 1991)

The SEMPA method has been applied in the study of domains and domain walls in a wide range of magnetic materials including recording media and recording heads; the reader is referred to the review articles for comprehensive lists. We illustrate the use of SEMPA with three examples. The first of these is taken from the work of the NIST group (Scheinfein et al 1991) in which they studied the remanent domain structure of the P2 layer of a conventional thin film recording head. Fig.17a,b show the resulting images with contrast sensitive to magnetisation in the direction of the arrows at the side of the images. It is seen that the large domains have antiparallel magnetisation and are separated by 180° walls. From Fig.17b it is seen that the closure domains have magnetisation along

the edge direction and that the sense of rotation of the surface magnetisation of the 180°
walls is not always the same. These images are comparable to these observed by Kerr
microscopy and have the advantage over the Type II SEM method of giving information
on the nature of the walls separating domains. To date, no experimental studies have
been reported in which the dynamic behaviour of head magnetisation was investigated.

Our second example arises from studies of the video recording characteristics of metal
evaporated tape (MET) and will also highlight the complementary role of the magnetic
imaging tools we are discussing. The MET studied comprised a 150nm film of $Co_{80}Ni_{20}$
evaporated on to aluminium-coated PET polymer. For the TEM/STEM studies, the film
had to be isolated from the tape. In video recording, two write heads are employed and
they are skewed through angles $\pm10°$ to the normal to the track direction which itself lies
at 5° to the tape direction. Given that the film has uniaxial anisotropy in the tape
direction, it is to be expected that the recording characteristics for adjacent tracks could
be different. A section of tape was first d.c. erased, i.e. magnetised along the tape
direction, and then low frequency tracks ($\lambda = 8.6\mu m$) were written. In TEM/STEM
studies, it proved very difficult to image the track written with the head field closest to
the easy axis direction, whereas the other track was easily visible but mainly through
contrast arising from induction changes in the direction normal to the track; this is
illustrated by the pair of DPC images in Fig.18a,b, where it can be seen that the zigzag

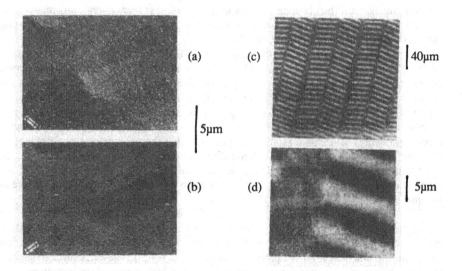

Figure 18 Comparison of SEMPA and DPC images from written tracks in $Co_{80}Ni_{20}$ MET tape for video
recording. (a),(b) DPC images from one of the tracks; the direction of magnetic induction to which the
contrast is sensitive is indicated by an arrow; (c),(d) SEMPA images with contrast sensitive to magnetisation
along the track direction. E=120keV (Oepen and Kirschner 1991, Sinclair 1990)

nature of the bit transition region is visible, but otherwise the contrast in the track vicinity is weak. The image sensitive to induction in the direction normal to the track shows much stronger contrast. Based on these observations, a model of the writing process was proposed (Ferrier et al 1987: Sinclair 1990). In this, the magnetisation in each track is initially in the direction dictated by the head geometry, but after the head has passed, it relaxes into or close to the direction of the easy axis. This implies that the magnetisation in the written bits will be similar for adjacent tracks. However, the stray field distribution established above/below the written bits will be determined by the geometry of the bits and will be different for adjacent tracks. Thus, for the track written closest to easy axis direction, we would expect in DPC imaging almost complete cancellation of the magnetisation by the stray field and hence very weak contrast. For the track imaged in Fig.18a,b, there would not be complete cancellation, and we should also see a net component of induction normal to the track direction and this would be antiparallel in adjacent bits. The results of the SEMPA study conducted by Oepen and Kirschner (1991) confirm this model. SEMPA images taken from a set of written tracks (note that although the track wavelength is the same, the bit width in this case was ~34μm compared to the case of Fig.18a,b where it was ~17μm) are shown in Fig.18c,d. There is of course no contrast arising from the stray field and the contrast is optimised for the component of magnetisation along the track. It is clear that we cannot differentiate between adjacent tracks on the basis of the contrast and this agrees with our model. Another feature of interest is that in the high magnification image Fig.18d we do not discern any zigzag nature to the domain (bit) boundary. This may indicate that there is rearrangement of the domain wall magnetisation close to the surface but this will require confirmation using samples written on the same video machine.

For our final example of the application of the SEMPA method we return to the work of the NIST group (Unguris et al 1990), who, in an attempt to understand the noise performance in recording, investigated different thin film media and two examples are shown in Fig.19. Fig.19a shows an image sensitive to magnetisation in the track direction for a $Co_{86}Cr_{12}Ta_2$ medium which exhibits good recording noise characteristics. This may be compared to a similar image (Fig.19b) for a $Co_{75}Ni_{25}$ medium which was of poorer recording quality. It seems reasonable to postulate that the poorer noise performance of this sample arises from the badly defined domain configuration in the written bits. In Fig.19c is shown the topographic image for the CoCrTa alloy specimen taken with the graphite detector; only after subtraction of this from the original magnetisation image is the quality of image shown in Fig.19b obtained.

The introduction of SEMPA as a tool for micromagnetic characterisation has undoubtedly proved a major advance and its potential for studies of magnetic recording media and transducers is considerable. Given the rather low efficiency of the polarisation detection, the practical limit to spatial resolution (30 - 40nm) is lower than is desirable, but nevertheless adequate for the many investigations which currently are important. The requirement of a UHV specimen environment is a disadvantage as is the need to remove any non magnetic surface layer, e.g. the carbon overcoat on thin film disks, prior to investigation.

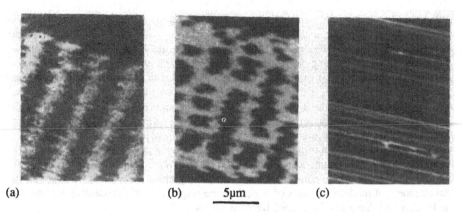

Figure 19 (a) SEMPA image sensitive to magnetisation in the track direcrion for a CoCrTa medium; (b) similar image to (a) but for a CoNi medium sample; (c) topographic contrast image for the same sample area as in (b). (Unguris et al 1990)

4. Kerr Optical Microscopy

The interaction of plane polarised light with a magnetic medium can lead to changes in the light which can usefully be used to reveal magnetic domain structure. It was Faraday who first proposed that plane polarised light, transmitted through a medium in the presence of a magnetic field with a component parallel to the direction of propagation, would be subject to a rotation in the plane of polarisation. The method was first applied to the study of ferromagnetic thin films by Fowler and Fryer (1954). The plane of polarisation of incident plane polarised light is also changed on reflection at a magnetised surface; this is the well known Kerr effect and it is the modern realisation and application of Kerr microscopy which we will discuss.

Kerr microscopy can be classified into several types depending on the relative orientation of the magnetisation, and the plane of incidence and direction of the electric vector of the incident light. The basics of a Kerr microscope comprise a light source, a sheet polariser to provide plane polarised light incident on the specimen and a rotatable sheet analyser to convert rotations of the polarisation plane into intensity variations in the image. The physics of optical reflection from metal surfaces is discussed in the textbooks on physical optics; here we note two aspects. Firstly, the depth of penetration of optical radiation into metals - the skin depth - is wavelength dependent and very small being of the order of nanometres, thus Kerr microscopy is essentially a technique for the exploration of surface magnetisation. Secondly, the interaction of polarised light with the specimen is weak and polarisation plane rotation is typically minutes of arc. For example, in permalloy the Kerr component of light reflected from the surface has an intensity of the order 10^{-6} of the reflected intensity (Argyle et al 1987). Crossed polarizer/analyser combinations do have the sensitivity to detect this, although not easily. Enhancement of the Kerr rotation angle would be an advantage and this can be achieved by a surface "blooming" technique introduced in the 1950s and now standard practice. This is achieved by evaporating a thin $\lambda/4$ dielectric film, usually zinc sulphide, on to the specimen surface and relies on interference effects in the layer. The major problem, however, with Kerr imaging is that unless the surface to be examined is very smooth and optically flat, Kerr contrast can be swamped by non-magnetic contrast from the specimen present under the crossed polariser/analyser condition. Fowler and Fryer(1954) attempted to overcome this difficulty by taking two images of the sample surface, one in the required state of magnetisation and the other with the specimen saturated by an external field; the latter should contain only non-magnetic information. If one image was a negative and the other a transparent positive then careful superposition gave a combined image in which non-magnetic contrast was reduced; the method was, however, only partially successful due to non-linearities in the photographic process. It is, however, the development of modern methods of digital image acquisition and processing (Schmidt et al 1985) which have revolutionised Kerr microscopy.

The modern Kerr microscope is usually built around a high performance commercial polarising microscope. The light source will be either a mercury arc, which is still preferred for some studies, or a laser. For high resolution studies the high quality objective lens is normally used in the oil immersion mode, giving a high numerical aperture. Illumination on the specimen is usually restricted by an aperture to about one quarter the diameter of the objective. If this aperture is centred then polar Kerr imaging is achieved. For conventional longitudinal Kerr imaging the aperture is displaced from the optic axis and the light is incident obliquely on the specimen. The analyser is normally in sheet form and is used in combination with a compensator which helps to reject that part of the background which is elliptically polarised. The image detection is usually performed by an image-intensified TV camera, the output of which is controlled by a video processor, supplemented by at least two frame stores. This allows images to be captured, digitised and stored in memory; to reduce the effect of random noise a selected number of successive images can be averaged by recursive filtering. The capability of subtracting images formed in different ways is central to the processing strategy and the possibility of carrying out operations such as image contrast expansion is normally available.

There are a number of different ways in which the Kerr microscope can be used to study static or quasi-static magnetisation distributions in a specimen. In all cases the interference effect of non-magnetic contrast present in the raw Kerr image must be removed. One method used is to acquire an image of the specimen surface in the magnetically saturated state which is achieved using a magnetic field from an external magnet in the specimen vicinity; in the case of thin film recording heads the same result may be achieved by exciting the head coil with a large d.c. current. This reference image is then subtracted from the Kerr contrast image(s). If we are to assess the magnetisation in the sample we need to record at least two images sensitive to magnetisation in orthogonal directions. By using both transverse and longitudinal polarised light we can achieve this without moving the specimen. This is illustrated by a pair of images in Fig.20, which were taken (Smith and Goller 1990) in longitudinal Kerr mode and processed as discussed above. In Fig.20b it should be noted that the long 180° walls are

169

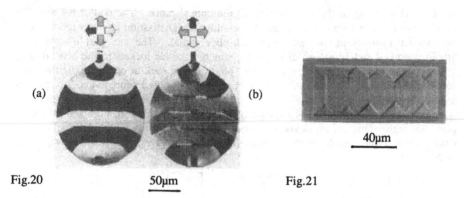

(a)　　　　　　　　　　　(b)

40µm

Fig.20　　　　　50µm　　　　Fig.21

Figure 20a,b Kerr micrographs for a permalloy layer patterned in the form of a polepiece. The arrows indicate the relation between magnetisation and Kerr contrast. (Smith and Goller 1990)
Figure 21 Kerr micrograph for a permalloy film sample. The image is the result of the subtraction of two images recorded with the film subjected to slightly different magnetic fields. (Argyle et al 1987)

made visible by the Néel character of the wall close to the surface. The change of contrast along the wall A shows the method is sensitive to the chirality of the wall and the presence of Bloch lines are also evident. Another method which has been used is to acquire two images with slightly different magnetic fields applied to the specimen. If the system is set to image domain contrast then the movement of domain boundaries will show up as black or white contrast and the extent of the contrast will be governed by the amount of movement. Given that the non-magnetic contrast is the same for each of the pair of raw images it will be absent in the subtracted image; an example of this procedure is shown in Fig.21 (Argyle et al 1987).

In studying the role of the recording head in determining the performance of a recording system it is important that information on the dynamic behaviour of the head magnetisation is investigated and this should be carried out at least at the frequency of recording and ideally at substantially higher values (well into the MHz range). Over the last few years a number of groups (for a review see Kryder et al 1990) have shown that Kerr microscopy is well suited to such studies and this may prove its greatest contribution. Two main approaches to dynamic investigations have been followed. In the first, which has become known as Scanning Kerr Effect Microscopy [SKEM] (Kasiraj et al 1986, Re and Kryder 1984), the spatial and temporal aspects of the magnetisation are separated by employing the technique of scanning optical microscopy (Kino and Corle 1989). An example of a SKEM apparatus due to Kryder and his co-workers is shown in Fig.22. The system utilises a polarising microscope configured in the confocal arrangement. The beam from the laser light source is expanded to a plane wave, passes through a polariser and is incident on the objective lens which produces a diffraction limited spot ~0.3µm in diameter on the specimen surface. After reflection the beam passes through an analyser and is focused into a measuring pin hole aperture and detected by a high speed photomultiplier. Imaging is achieved by precision two dimensional

170

mechanical scanning of the specimen under the optical probe. The coil of the thin film head is driven with an a.c. signal and the resulting magnetisation changes are reflected in intensity changes in the raw photomultiplier signal. The a.c. part of this, after amplification, is fed to a lock-in amplifier which is phase locked to the head driving voltage. The r.m.s. average amplitude and phase of the lock-in output is digitised and stored on computer for the point by point scan. The d.c. component of the raw photomultiplier signal relates to the reflectivity of the local surface and is recorded simultaneously. The method is capable of high spatial resolution, high sensitivity for periodic magnetisation changes and the bandwidth is limited only by the detection system (lock-in amplifiers suitable for this application are currently available with bandwidths up to 50MHz).

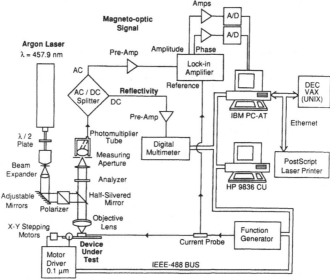

Figure 22 Schematic diagram of a scanning Kerr magneto-optic photometer. (Kryder et al 1990, c IEEE)

We illustrate the use of this scanning photometry method with two examples. Kasiraj et al (1986) were the first group to study by imaging the magnetisation dynamics in the P2 polepiece of a thin film head. Kryder and his group (Kryder et al 1990) have also exploited the technique for the same purpose and an example from their work is given in Fig.23 which shows the longitudinal Kerr image for a P2 polepiece driven at 1MHz. The contrast which is present is due both to domain motion and magnetisation rotation, the latter being clearly visible near the back closure. Domain walls will only be delineated if the wall motion is greater than the spatial resolution and the extent of the observed contrast is related to the distance over which a wall moves during the period of excitation. The image shows that at 1MHz, flux is conducted both by domain wall motion and by magnetisation rotation; the relative importance of these mechanisms as a function of frequency is easily investigated by the technique. Kasiraj et al (1987) introduced an interesting variation of the SKEM method to provide data on the time evolution of the magnetisation. In this case they excited the head with an a.c. voltage of sufficient magnitude to saturate the head. The lock-in amplifier was replaced by a digital oscilloscope with a 1.0ns time resolution, and this allowed the signal waveform to be recorded over the drive period and averaged to improve signal/noise. Time portions of the signal could then be extracted and used to give a series of images representing the time evolution of the magnetisation.

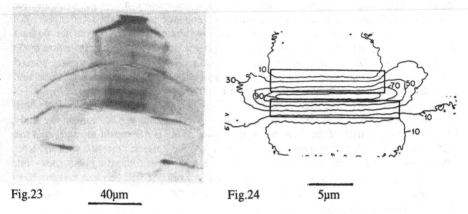

Fig.23   40μm   Fig.24   5μm

Figure 23 Image of Kerr magneto-optic response for a P2 polepiece driven at 120mA-turn at 1MHz. Note the contrast in the region of the closure indicating a large change in magnetisation direction during the excitation period. (Kryder et al 1990, c IEEE)

Figure 24 Contours of rms longitudinal Kerr signal and hence magnetisation from a thin CoNi film placed 0.25μm above the air bearing surface of a thin film head. (Koeppe et al 1988)

The second example we cite is from the work of Koeppe et al (1988) and this involved a study of the magnetisation of the poletips of a recording head. Using SKEM the dynamic magneto-optic response of the outward normal component of magnetisation at the air bearing surface was imaged. They were able to demonstrate that for a thin film head with asymmetric polepieces, the switching of the magnetisation showed distinct variations both across and along the polepieces. The authors were able to extend significantly the useful information from this study by introducing a thin CoNi film evaporated on to a glass coverslip. The film, which had an in-plane easy axis of magnetisation, was held ~0.25μm distant from the polepieces, i.e. at a typical flying height, and the magnetisation in the film was monitored by the SKEM method using the longitudinal magneto-optic contrast mode. The results map the head field in the film plane and a contour map for the head is shown in Fig.24 and clearly demonstrates the side-writing asymmetry which would be present in tracks written with the head.

The second method of Kerr imaging for the study of dynamic effects utilises the technique of stroboscopic imaging to provide instantaneous observation of the magnetisation process (Kryder et al 1990). The light source is a pulsed laser and wide field Kohler illumination is employed to image the whole specimen surface

172

simultaneously. The polariser/analyser arrangement is standard and the longitudinal mode of Kerr imaging is normally used. The images are captured on a silicon-intensified-target (SIT) camera. Dynamic observations are made by synchronising the 10ns laser pulses with the drive to the head coil and the frame rate of the camera; this is controlled using a delay generator. The normal methods of image processing used in Kerr microscopy are available. One method of image normalisation which has not been mentioned previously is the use of a "grey" image to be used in image subtraction for non-magnetic contrast removal. This image is obtained by using a very large amplitude a.c. drive voltage in the head coil which causes the magnetic contrast to be smeared out over the head image. An example of the use of this normalisation method is shown in Fig.25, where the domain magnetisation for a P2 layer is shown at two different points in the magnetisation cycle performed at 1MHz. An alternative approach is to record two images at closely spaced times in the sinusoidal magnetising cycle and subtract them; the positions of walls and the extent of their movement are then visible in the processed image.

In this section we have been concerned with the performance of Kerr imaging in both its static/quasi-static and dynamic modes and have illustrated its applications in the study of the micromagnetism of thin film recording heads. This is not meant to imply that the method is not of value in studying media - in practice the greatest current use of the Kerr effect is as the detection method in commercial magneto-optic recording disk files - but rather the present emphasis of the new experimental work. It will be clear that the Kerr microscope has already proved its value and in research laboratories it is the most common microstructural tool

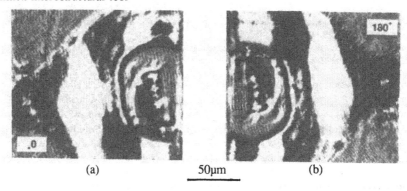

(a)   50µm   (b)

Figure 25 Kerr image of the magnetisation state of a thin film recording head at (a) 0° and (b) 180° of the phase of a sinusoidal 1MHz 320mA-turn p-p drive current. (Kryder et al 1990, c IEEE)

5. Magnetic Force Microscopy

In the field of the microscopy of materials the most important advance in the 1980s was undoubtedly the introduction of the scanning tunnelling microscope [STM] (Binnig and Roher 1985) and the instruments which it has spawned such as the atomic force microscope [AFM] (Binnig et al 1986) and, of particular importance to this discussion, the magnetic force microscope [MFM] (Martin and Wickramasinghe 1987). The novel feature of the STM is that structural features of the specimen are revealed through the interaction between a sharp needle - the tip - and a flat electrode - the specimen; the tunnelling signal is strongly dependent on the separation between these two conductors. In practice the image is created by scanning the sample under the tip and the method is capable of high sensitivity to height changes ~1.0pm and of high spatial resolution ~0.2nm. The introduction of the STM represented a substantial technological achievement, but once accomplished, its variants followed fairly rapidly and in all cases the same idea of monitoring the interaction between a tip and the sample is central to the operation. In the case of the AFM the tip is mounted on a very sensitive cantilever and the force or force gradient between sample and tip is monitored; in the case of the MFM the tip is ferromagnetic and the interaction of concern is that arising from the stray magnetic field distribution above the specimen surface. A short review by Martin et al (1990) serves as a useful introduction to the design, operation and applications of the MFM.

A schematic diagram of a basic MFM is shown in Fig.26. The essential feature is the sharp ferromagnetic tip either mounted on or part of a soft cantilever, which on bending moves the tip towards or away from the specimen surface; typical specimen/tip separation is in range 10 - 500nm. The cantilever is mounted on a piezoelectric bimorph which can be used to control the static height of the probe and/or to oscillate it with an amplitude in the range 1.0 - 10.0nm. The magnetic interaction strength is determined using a high sensitivity sensor normally based either on optical interferometry or tunnelling. The sample is mounted on a stage whose xy position can be controlled by

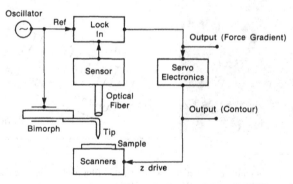

Figure 26 Block diagram of a magnetic force microscope. (Rugar et al 1990)

piezoelectric sensors and they are excited to scan the specimen below the tip in a suitably controlled manner. The most obvious way of sensing the interaction is to determine the magnetic force on the tip, which can be attractive or repulsive, by measuring directly the deflection of the cantilever. The resulting image contains information on the local stray magnetic field. High spatial resolution can be achieved with this method. Grutter et al (1988), using a tunnelling sensor, studied the domain wall transitions in a NdFeB and claimed spatial resolution at the 10nm level. This form of measurement can be influenced by thermal drift and a method which overcomes this problem is to measure the force gradient sensed by the tip rather than the force itself. In this case the cantilever is vibrated at a frequency just above its resonant frequency, which is typically in the range 10 - 100kHz. Variations of the force on the tip, which includes a Van der Waals contribution, changes the effective spring constant of the cantilever and shifts its resonant frequency. This is reflected in a small change in the tip oscillation amplitude and it is this which is sensed and incorporated in a feedback loop to adjust the signal and alter the Z piezo to maintain the oscillation amplitude constant. The Z piezo signal is used to generate the image; provided the cantilever is sufficiently weak to reduce the influence of topography, the variations in contrast are dominated by magnetic effects. Results of interest to magnetic recording investigations have now been obtained with this imaging method and as an example we show some images from the work of Rugar et al (1990) who investigated thin film longitudinal recording disks. Fig.27 shows the image of 5μm spaced bits in a 50nm thick CoPtCr alloy medium. This was recorded using a tip oriented normal to the specimen surface and reflects the distribution of the normal component of the stray field. Higher magnification images of the domain boundaries do show intensity variations along the wall, but it is hard to relate this to the type of transition observed in high resolution (S)TEM studies for media of similar magnetic characteristics. It should be noted that the nature of the contrast observed in the MFM is dependent on the orientation of the tip axis and hence its magnetisation relative to the specimen surface.

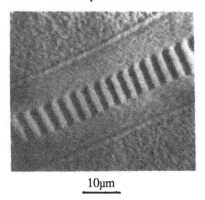 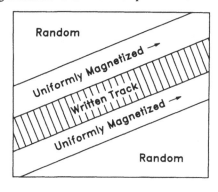

10μm

Figure 27 MFM image from a track written onto a CoPtCr thin film medium in a dc-erased state. The axis of the tip was normal to the specimen. (Rugar et al 1990)

The third mode of operation for the MFM is its use to study time-dependent magnetic stray field phenomena arising from the specimen itself or induced by the ferromagnetic tip. In Fig.28 we see the image of the magnetic field beyond the poletips of a thin film recording head (Martin et al, 1989); here the head coil is energised by an a.c. voltage and the resulting stray field induces an a.c. oscillation of the tip which is detected as discussed above. A variant on this approach is to oscillate the tip above the polepiece and study the variation of magnitude of the voltage induced in the head coil for relative motion between the tip and the sample.

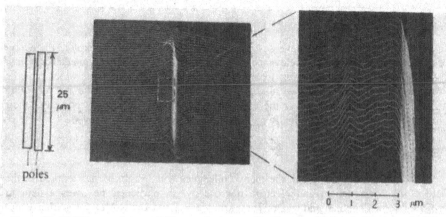

Figure 28 MFM image of the ac field strength above a thin film recording head. (Martin et al 1989)

The simplest interpretation of an MFM image would assume that the tip can be represented by a point dipole and that the dipole has no effect on the magnetisation of the sample or vice versa; i.e. this implies the tip and sample are both magnetically hard. In addition, the complication of interpreting topography contrast should be avoided by using smooth surfaces. For thin film recording media with smooth surfaces this dipole approximation is probably reasonable for spatial resolution at the tens of nanometre level. In general, however, the shape of the tip, the nature of its magnetic structure and the correct accounting for induced magnetisation changes in sample and/or tip will have to be taken into account in determining the function required for a deconvolution procedure. The importance of accurate deconvolution methods will increase as the required spatial resolution increases. One interesting feature of MFM for studies of magnetic recording arises through the possibility of using the tip to write "bits" on to a magnetic medium and then use the same tip to interrogate the micromagnetic structure of the written information. The "write" process would be performed with the tip very close to the surface and the "read" function with the tip moved back - already one such study has been reported (Moreland and Rice, 1990).

Magnetic force microscopy has already shown its usefulness for the study of magnetic recording media and inductive transducers; the possibility of studying dynamic effects is also established. The great advantage of the method is that the specimen preparation is minimal and presence of overlayers, such as the carbon overlayer on modern thin film disks, is not a problem. Whilst high spatial resolution has been demonstrated, the extraction of meaningful data on the stray field distribution of the specimen will require deconvolution techniques and it remains to be established how effective these will prove. As part of the evaluation process of MFM and its development over the next few years it will be important to carry out combined studies where the same sample is investigated

176

by as many other techniques as possible, particularly those capable of high spatial resolution.

6. Conclusion

In the foregoing sections we have described experimental methods to explore the nature of micromagnetic structure of relevance to the performance of digital magnetic recording. It should now be clear that no one technique can answer all the questions which must be posed and this is true even if we subdivide our considerations into media and recording heads. In combination however the experimental methods are already powerful; their ultimate potential, both individually and collectively, remains to be proved.

Acknowledgements

The author wishes to thank the many colleagues with whom he has worked in the field of micromagnetism and electron microscopy. In particular he owes a debt to Professor John Chapman and to Drs Tom Arnoldussen and Denis Mee for guidance and many helpful discussions.

References

Abraham, D.L. and H.Hopster, "Magnetic Probing Depth in Spin Polarised Secondary Electron Spectroscopy", *Phys. Rev. Lett.*, **58**, 1352 (1987)

Aharonov, Y. and D.Bohm, "Significance of Elecromagnetic Potentials in the Quantum Theory", *Phys. Rev.,* **115**, 485 (1959)

Alexopoulos, P.S. and R.H.Geiss, private communication (1985)

Allenspach, R., M.Taborelli and M.Landolt, "Oxygen On Fe(100): an Initial-Oxidation Study by Secondary-Electron Spectroscopy", *Phys Rev Lett*, **55**, 2599 (1987)

Argyle, B.E., B.Petek and D.A.Herman, "Optical Imaging of Magnetic Domains in Motion", *J. Appl. Phys.*, **61**, 4303 (1987)

Arnoldussen, T.C., E.M.Rossi, A.Ting, A.Brunsch, J.Schneider and G.Trippel, "Obliquely Evaporated Iron-Cobalt and Iron-Cobalt-Chromium Thin Film Recording Media", *IEEE Trans. Mag.*, **MAG-20**, 821 (1984)

Arnoldussen, T.C., this volume (1992)

Banbury, J.R. and W.C.Nixon, "A High-Contrast Directional Detector for the Scanning Electron Microscope", *J. Phys. E*, **2**, 1055 (1969)

Binnig, G., C.F.Quate and C.Gerber, ""Atomic Force Microscope", *Phys. Rev. Lett.*, **56**, 930 (1986)

Binnig, G. and H.Rohrer, "The Scanning Tunnelling Microscope", *Sci. Amer.,* **253**, 50 (1985)

Bitter, F., "Inhomogeneities in the Magnetisation of Magnetic Materials", *Phys. Rev.,* **38**, 1903 (1931)

Chapman, J.N., E.M.Waddell, P.E.Batson and R.P.Ferrier, "The Direct Determination of Magnetic Domain Wall Profiles by Differential Phase Contrast Electron Microscopy", *J. Ultramicroscopy*, **4**, 283 (1979)

Chapman, J.N., "The Investigation of Magnetic Domain Structures in Thin Foils by Electron Microscopy", *J. Phys. D*, **17**, 623 (1984)

Chapman, J.N., I.R.McFadyen and J.P.C.Bernards, "Investigation of Cr Segregation Within rf-sputtered CoCr Materials", *J. Magn. Mag. Matls.*, **62**, 359 (1986)

Chapman, J.N., I.R.McFadyen and S.McVitie, "Modified Differential Phase Contrast Lorentz Microscopy for Improved Imaging of Magnetic Structures", *IEEE Trans. Mag.*, **MAG-26**, 1506 (1990)

Chapman, J.N. and I.R.McFadyen, "Electron Microscopy of Magnetic Materials", to be published in *EMSA Bulletin*, Spring-1992

Chrobrok, G. and M.Hoffmann, "Electron Spin Polarization of Secondary Electrons Ejected from Magnetised Europium Oxide", *Phys. Lett.*, **57a**, 257 (1976)

Elsbrock, J.B. W.Schroeder and E.Kubalek, "Evaluation of Three-Dimensional Micromagnetic Stray Fields by Means of Electron Beam Tomography", *IEEE Trans. Mag.*, **MAG-21**, 1593 (1985)

Ferrier, R.P., "Small Angle Electon Diffraction", in *Advances in Optical and Electron Microscopy*, **Vol 3**, ed. R.Barer and V.E.Cosslett, (New York: Academic Press), 155 (1969)

Ferrier, R.P. and D.F.Kyser, "Magnetic Contrast in the Secondary Emission Mode of the SEM", *Proc. VIIIth MAS Conf. New Orleans*, 49 (1973)

Ferrier, R.P., H.C.Tong, K.Parker and R.H.Geiss, "Lorentz Microscopy of a Thin Film High Density Magnetic Recording Medium", *Inst. of Physics Conf. Ser.*, **68**, 193 (1983)

Ferrier, R.P. and R.H.Geiss, "The Study of Magnetic Domains in Thin Film Recording Heads", *Proc. XIth Int. Conf. on Electron Microscopy, Kyoto*, **2**, 1725 (1986)

Ferrier, R.P., R.H.Geiss and P.S.Alexopoulos, "The Study of Written Tracks in Digital Magnetic Recording Media", *Proc. XIth Int. Cong. on Electron Microscopy, Kyoto*, **2**, 1727 (1986)

Ferrier, R.P., M.D.Ferry and J.L.S.Wales, "Studies of the Micromagnetic Structure of Recorded Tracks in MET", *IEEE Trans. Mag.*, **MAG-22**, 3639 (1987)

Ferrier, R.P., F.J.Martin, T.C.Arnoldussen, and L.L.Nunnelley, "Lorentz Image-Derived Film Media Noise", *IEEE Trans. Mag.*, **MAG-24**, 2709 (1988)

Ferrier, R.P., F.J.Martin, T.C.Arnoldussen, and L.L.Nunnelley, "The Determination of Transition Noise by Lorentz Electron Microscopy", *IEEE Trans. Mag.*, **MAG-25**, 3357 (1989)

Ferrier, R.P., F.J.Martin, T.C.Arnoldussen, and L.L.Nunnelley, "An Examination of Transition Noise by Lorentz Electron Microscopy", *IEEE Trans. Mag.*, **MAG-26**, 1536 (1990a)

Ferrier, R.P., S.McVitie and W.A.P.Nicholson, "Magnetization Distributions in Thin Film Recording Heads by Type II Contrast", *IEEE Trans. Mag.*, **MAG-26**, 1337 (1990b)

Ferrier, R.P. and S.McVitie, "A New Method for the Observation of Type II Magnetic Contrast", *Proc. XIIth Int. Cong. on Electron Microscopy, Seattle*, **4**, 764 (1990)

Ferrier, R.P. and S.McVitie, "The Depth Sensitivity of Type II Magnetic Contrast", *Proc. EMSA - 49th Annual Meeting, San Jose*, 766 (1991)

Fowler, C.A. and E.M.Fryer, "Magnetic Domains in Cobalt by the Longitudinal Kerr Effect", *Phys. Rev.*, **95**, 564 (1954)

Gordon, R., "A Tutorial on ART", *IEEE Trans. Nucl. Sci.*, **NS-21**, 78 (9174)

Hale, M.E., H.W.Fuller and H.Rubenstein, "Magnetic Domain Observation by Electron Microscopy", *J. Appl. Phys.*, **30**, 789 (1959)

Grutter, P., E.Meyer, H.Heizelmann, H.Rosenthaler, H.-J.Hedber and H.-S.Guntherodt, "Application of Atomic Force Microscopy to Magnetic Materials", *J. Vac. Sci. Tech.*, **A6**, 279 (1988)

Ikuta, T. and R.Shimizu, "Analysis of Type-II Magnetic Contrast from Ferromagnetic Thin Films in the Scanning Electron Microscope", *J. Magn. Mag. Matls.*, **35**, 356 (1983)

Kasiraj, P., R.M.Shelby, J.S.Best and D.E.Horne, "Magnetic Domain Imaging with a Scanning Kerr Effect Microscope", *IEEE Trans. Mag.*, **MAG-22**, 837 (1986)

Kasiraj, P., D.E.Horne and J.S.Best, "A Method for the Magneto-optic Imaging of Magnetisation Time Evolution in Thin Films", *IEEE Trans. Mag.*, **MAG-23**, 2161 (1987)

Kino, G.S. and T.R.Corle, "Confocal Scanning Optical Microscopy", *Physics Today*, **42**, 52 (1989)

Koeppe, P.V., M.E.Re and M.H.Kryder, "Effect of Poletip Alignment on Magnetic Fringing Fields from Recording Heads" *J. Appl. Phys.*, **63**, 4042 (1989)

Kryder, M.H. P.V.Koeppe and F.H.Liu, "Kerr Effect Imaging of Dynamic Processes", *IEEE Trans. Mag.*, **MAG-26**, 2995 (1990)

McVitie, S. and R.P.Ferrier, "Model Stray Field Calculations of a Longitudinal Recording Medium", to be published in *Proc XIIIth Int. Cong. on Magnetism, Edinburgh 1991.*

Martin, Y. Abraham D.W. Hobbs P.C.D. and Wickramasinghe H.K., "Magnetic Force Microscopy - a Short Review", *Electrochemical Society Proc. on Magnetic Materials, Processes and Devices*, **90-98**, 115 (1989)

Martin, Y. and H.K.Wickramasinghe, "Magnetic Imaging by 'Force Microscopy' with 1000Å Resolution", *Appl. Phys. Lett.*, **50**, 1455 (1987)

Matsuda, J., K.Aoyagi, Y.Kondoh, M.Iizuka and K.Mukasa, "A Three-Dimensional Measuring Method for Magnetic Stray Fields", *IEEE Trans. Mag.*, **MAG-26**, 2061 (1990)

Mee, P.B., "SEM Observationss of Domain Configurations in Thin Film Head Pole Structures", *J. Appl. Phys.*, **51**, 861, (1980)

Moreland, J. and P.Rice, "High-Resolution, Tunnelling-Stabilised Magnetic Imaging and Recording", *Appl. Phys. Lett.*, **57**, 310 (1990)

Morrison, G.R. and J.N.Chapman, "A Comparison of Three Differential Phase Contrast Systems Suitable for Use in STEM", *Optik*, **64**, 1, (1983)

Munro, E., "Computer-Aided-Design Methods in Electron Optics", *Ph.D. Dissertation*, University of Cambridge (1971)

Oepen, H.P. and J.Kirschner, "Imaging of Magnetic Microstructures at Surfaces: the Scanning Electron Microscope with Polarisation Analysis", *Scanning Microscopy*, **5**, 1 (1991)

Re, M.E. and M.H.Kryder, "Magneto-Optic Investigation of Thin-Film Recording Heads", *J. Appl. Phys.*, **55**, 2245 (1984)

Rogers, D.J., J.N.Chapman, J.P.C.Bernards and S.B.Luitjens, "Determination of Local Composition in CoCr Films Deposited at Different Substrate Temperatures", *IEEE Trans Mag.*, **MAG-25**, 4180 (1989)

Rugar, D., H.J.Mamin, P.Guthner, S.E.Lambert, J.E.Stern and I.R.McFadyen, "Magnetic Force Microscopy: General Principles and Applications to Magnetic Recording", *J. Appl. Phys.*, **68**, 1169 (1990)

Scheinfein, M.R., J.Unguris, M.H.Kelley, D.T.Pierce and R.J.Celotta, "Scanning Electron Microscopy With Polarisation Analysis (SEMPA)", *Rev. Sci. Instrum.*, **61**, 2501 (1990)

Schmidt, F., W.Rave and A.Hubert, "Enhancement of Magneto-Optical Domain Observation by Digital Imaging Processing",", *IEEE Trans. Mag.*, **MAG-21**, 1596 (1985)

Seiler, H.J., "Secondary Electron Emission in the Scanning Electron Microscope", *J. Appl. Phys.*, **54**, R1 (1983)

Smith, A.B. and W.W.Goller, "New Domain Configurations in Thin-Film Recording Heads", *IEEE Trans. Mag.*, **MAG-26**, 1331 (1990)

Steck, M., H.Schewe and E.Kubalek, "Three-Dimensional Magnetic Field Measurement of a Modified Thin Film Magnetic Head for Vertical Recording by Means of Lorentz Tomography", *IEEE Trans. Mag.*, **MAG-26**, 1343 (1990)

Tonomura, A., "Electron Holography", *Proc. XIth Int. Cong. on Electron Microscopy*, **Vol 1**, 9 (1988)

Tsuno, K., "Magnetic Domain Observation by Means of Lorentz Electron Microscopy with Scanning Technique", *Rev. Solid State Science*, **2**, 1 (1988)

Unguris, J., D.T.Pierce and R.J.Celotta, "Scanning Electron Microscopy with Polarisation Analysis: Studies of Magnetic Microstructure", *Rev. Sci. Instrum.*, **57**, 1314 (1986)

Unguris, J., M.R.Scheinfein, R.J.Celotta and D.T.Pierce, "Scanning Electron Microscopy with Polarisation Analysis: Studies of Magnetic Structures", in *Chemistry and Physics of Solid Surfaces VIII*, ed R.Vanselow and R.Howe, (Springer-Verlag, Germany), 239, (1990)

Wells, O.C. and R.J.Savoy, "Magnetic Domains in Thin Film Recording Heads as Observed in the SEM by a Lock-In Technique", *IEEE Trans. Mag.*, **MAG-17**, 1253, (1981)

Wells, O.C., "Some Theoretical Aspects of Type I Magnetic Contrast in the Scanning Electron Microscope", *J. of Microsc.*, **139**, 187 (1985)

CHAPTER 6

MICROMAGNETIC MODELING OF THIN FILM RECORDING MEDIA

Jian-Gang Zhu

Department of Electrical Engineering and
The Center for Micromagnetics and Information Technologies
University of Minnesota
Minneapolis, Minnesota 55455

1. Introduction

Thin metallic film media have become the leading recording media for high density recording applications, mainly due to its large saturation moment and high coercivity. In the past five years, longitudinal thin film recording media have undergone a significant advancement in reducing medium noise and increasing film coercivity. Remarkable areal recording densities of 1 $Gbits/in^2$ and 2 $Gbits/in^2$ have been achieved in laboratory demonstrations with significant signal to noise ratio (Yogi et al. 1990, Futamoto et al. 1991) and it is only a matter of time before they are realized in products. One of the main reasons for such rapid advances is the improvement in understanding the correlations between film microstructures and recording performance and noise properties gained through extensive experimental and theoretical studies.

The fundamental aspect of the microstructure of a thin film recording medium is its granularity (Arnoldussen 1986). These films are polycrystalline with magnetic crystallites arranged in the film with nearly unity packing. Almost all the films are Co-base alloys with saturation magnetization ranging from 300 to 800 emu/cm^3. Films used as practical recording medium usually have coercivity around 700 to 1500 Oe. It has been found experimentally that by varying deposition conditions, film coercivity and other magnetic hysteresis behavior can take on a broad range of values without even varying material composition. Film magnetic hysteresis properties are strong functions of film microstructures. With a film considered as an assembly of closely packed interactive magnetic grains, the hysteresis properties not only depend on the intrinsic magnetic properties of each individual magnetic grains, but also the magnetic interactions, including both magnetostatic and exchange interactions, among or in-between the grains in the film. Understanding of the correlations of film magnetic properties and recording performance with film microstructure and material magnetic parameters is critically important for the development of advanced recording media. Computer modeling combined with experimental studies can provide, and has been providing, us such needed understanding as well as the understanding of the underlying physics governing the magnetization processes in these films.

Magnetization reversal mechanisms of isolated single domain magnetic particles have been studied extensively with the development of micromagnetics theory (Stoner and Wohlfarth 1948, Shtrikman 1960, Brown 1963 and 1968) In thin film recording media, not only the micromagnetic behavior of a single particle is important, so are the magnetic interactions among the magnetic grains in the film. The interactions in these films are usually so significant that reversal mechanisms of isolated individual grains can be altered (Zhu and Bertram 1989). It is often the case that in these films, collective magnetization behavior dominates during the magnetization processes.

Hughes proposed a computer simulation model for studying magnetization reversal process in CoP thin film media (Hughes, 1983). In his model, an energy minimization method was used to obtain magnetization configurations during magnetization processes. Using the model, Hughes studied effects of magnetostatic interactions in the film to the hysteresis properties and collective magnetization reversals of the grains in the film. Victora studied the magnetization reversal process in obliquely evaporated CoNi films by solving Landau-Lifshitz-Gilbert dynamic equations (Victora 1987a and 1987b). Zhu and Bertram developed a theoretical model for studying magnetic thin films with granular structures in general (Zhu and Bertram 1988a). In Zhu and Bertram's model, assuming each grain is always uniformly magnetized, a film is considered as an assembly of interactive magnetic grains with inclusion of both long range magnetostatic interaction and nearest neighbor intergranular exchange coupling. Magnetization directions of the grains during a magnetization process are determined by solving coupled Landau-Lifshitz equations. Applying the model, Zhu and Bertram went on to carry out a rather thorough study of the fundamental magnetization processes in thin film recording media, including both longitudinal and perpendicular films. Mansuripur and Giles have implemented a similar micromagnetic model on the Connection Machine to study magnetization process in both magnetic and magneto-optical recording media (Mansuripur and Giles 1990). Miles and Middleton also proposed a similar simulation model, but adding additional features such as irregular spatial arrangement of spherical magnetic grains (Miles and Middleton 1990).

This review focuses on the simulation studies of the fundamental magnetization processes, recording performance and noise properties in thin film recording media. Only the longitudinal films are discussed here, although simulation studies has shown very interesting findings for the magnetization process in CoCr perpendicular films (Zhu and Bertram 1989 and 1991b). The review is mainly based on a series of studies conducted by Zhu and Bertram. Emphasis is placed on the effects of magnetostatic interaction and intergranular exchange coupling in the films. Description of the theoretical model for thin film recording media is given in Sec. 2. In Sec. 3, the magnetization reversal process and resulting hysteresis properties in longitudinal thin film media are reviewed. Noise behavior of recorded isolated transitions and di-bit transition pairs are discussed in Sec. 4. In Sec. 5, a much simplified theoretical approach and some of the important results are reviewed. In Sec. 6, study of reverse *dc* erase noise utilizing a simplified model and recent experimental investigations are discussed. In the text, *cgs* units are used.

2. Theoretical Model

The model considers a thin film, either longitudinal or perpendicular, as a monolayer of closely packed magnetic grains. The main assumption in the model is that each magnetic grain is always uniformly magnetized such that during magnetization processes, only its magnetization direction is changing. Since almost all the films utilized as recording media are *Co*-based magnetic alloys and the sizes of the grains in a typical recording film are in the range of 150 – 500Å in diameter, this assumption is rather reasonable. Such grain sizes are well below the critical size for the existence of multidomain behavior (for *Co*, the critical diameter of a multidomain sphere is about 1400Å). With this assumption, the problem reduces to the modeling of an assembly of *interacting* Stoner-Wohlfarth particles.

2.1. Modeling Array

A thin film is modeled by a two dimensional array of hexagonally shaped grains arranged on a triangular lattice, as shown in Fig. 2.1. In the figure, D is the grain diameter, d is the intergranular boundary separation and δ is the grain height which is also the film thickness. This array of hexagons was first used by Hughes to study CoP longitudinal films (Hughes 1983). The important features of this array include the following: (1) Each hexagon has six nearest neighboring grains, closely resembling the grain arrangement in a real film. It is critical to have number of nearest neighboring grains close to real films for obtaining correct magnetization configurations; (2) Intergranular boundary separation is uniform throughout the film; (3) Unity packing fraction can be obtained; and (4) A fast Fourier transformation method can be utilized for the calculation of the long-range magnetostatic interactions. In the model, each grain is assumed to be a single crystal. In this review, we will consider only crystal grains with uniaxial crystalline anisotropy, although other types of crystalline anisotropy, such as cubic anisotropy, can be included in the model calculation. The crystalline anisotropy easy axis of each grain can be oriented in any desired directions.

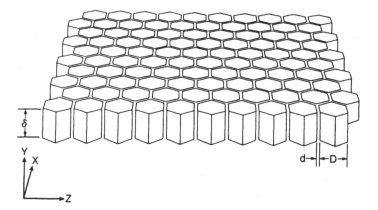

Figure 2.1 Illustration of 2D array of hexagons on a triangular lattice. δ is the grain height equal to the film thickness, D is the grain surface-to-surface diameter in the film plane and d is the intergranular boundary separation.

2.2. Energy Consideration

The free energy considered in this model includes the uniaxial crystalline anisotropy energy, the magnetostatic energy, the intergranular exchange energy and the Zeeman energy.

(1) *Crystalline anisotropy energy*. The uniaxial crystalline anisotropy energy density for a single crystallite (the ith grain in the array) is

$$E_{ani}(\mathbf{r}_i) = K \sin^2\theta_i, \quad i = 1, 2,..., N \tag{2.1}$$

where K is the anisotropy energy constant, θ is the angle between the easy axis orientation and the magnetization direction and N is the total number of grains in the array. In the model, it is rewritten in a vector form:

$$E_{ani}(\mathbf{r}_i) = K \, |\mathbf{k}_i \times \mathbf{m}_i|^2$$
$$= K\left[1 - (\mathbf{k}_i \cdot \mathbf{m}_i)^2\right] \quad i = 1, 2,..., N \tag{2.2}$$

where $\mathbf{k}$ is the unit vector along the crystalline easy axis, and $\mathbf{m}$ is the unit vector in the magnetization direction.

(2) *Magnetostatic interaction energy*. In the array, each grain interacts with all the other grains through magnetostatic interactions. For the ith grain, the magnetostatic interaction energy density averaged over the grain volume is written as:

$$E_{mag}(\mathbf{r}_i) = -\mathbf{M}_i \cdot \left[\sum_{j \neq i} D_{ij} \cdot \mathbf{M}_j + \frac{1}{2} D_{ii} \cdot \mathbf{M}_i \right] \qquad i = 1, 2, ..., N \qquad (2.3)$$

where the summation is over all the grains in the entire array and D_{ij} is the 3×3 interaction matrix. D_{ij} only depends on the geometric parameters of the array and is derived by integrating the magnetic poles on the surfaces of the jth grain and averaging over the volume of the ith grain, given as

$$D_{ij} = \frac{1}{v_i} \int_{v_i} dr^3 \int_{s_j} dr'^2 \frac{(\mathbf{r}-\mathbf{r}')\hat{n}'}{|\mathbf{r}-\mathbf{r}'|^3}. \qquad (2.4)$$

where $\mathbf{r}'$ is the position vector of the position on the surface of the jth grain and $\mathbf{r}$ is the position vector within the ith grain. The self-demagnetization energy is included in the D_{ii} term. Since only the magnetization directions are to be determined, the self-demagnetization energy only gives the shape anisotropy. For avoiding possible instability in the calculations caused by the component of the self-demagnetization field opposite to the magnetization direction, the term D_{ii} should be rewritten as a proper shape anisotropy form (Zhu 1989).

(3) *Intergranular exchange coupling*. In order to describe possible exchange coupling between nearest neighboring grains through grain boundaries, an intergranular exchange coupling was introduced in a form analogous to the spin exchange energy:

$$E_{exc}(\mathbf{r}_i) = -\frac{2A^*}{M^2 a^2} \mathbf{M}_i \cdot \sum_{n.n.} \mathbf{M}_j, \qquad i = 1, 2, ..., N \qquad (2.5)$$

where A^* is an effective exchange energy constant, M is the magnetization of each individual grain and a is the center-to-center distance between adjacent grains. The summation is over all the nearest-neighbor grains. The effective exchange energy constant A^* measures the exchange coupling between the adjacent grains and the Eq. 2.5 should be considered as a phenomenological approximation.

(4) *Zeeman energy*. This energy term is due to an externally applied field and is often referred to as magnetic potential energy:

$$E_{ext}(\mathbf{r}_i) = -\mathbf{H} \cdot \mathbf{M}_i, \qquad i = 1, 2, ..., N \qquad (2.6)$$

The total energy density of the ith grain is

$$E_{tot}(\mathbf{r}_i) = E_{ani}(\mathbf{r}_i) + E_{mag}(\mathbf{r}_i) + E_{exc}(\mathbf{r}_i) + E_{ext}(\mathbf{r}_i). \qquad (2.7)$$

The effective magnetic field on the ith grain is defined as the differentiation of the energy density with respect to the magnetization:

$$\mathbf{H}_i = -\frac{\partial E_{tot}(\mathbf{r}_i)}{\partial \mathbf{M}_i}$$

$$= -\frac{\partial E_{tot}(\mathbf{r}_i)}{\partial M_i^x}\mathbf{e}_x - \frac{\partial E_{tot}(\mathbf{r}_i)}{\partial M_i^y}\mathbf{e}_y - \frac{\partial E_{tot}(\mathbf{r}_i)}{\partial M_i^z}\mathbf{e}_z. \tag{2.8}$$

where $\mathbf{e}_x$, $\mathbf{e}_y$, and $\mathbf{e}_z$ are unit vectors along three axes of the coordinates x, y, and z. Since the crystalline anisotropy field effectively acts like a local constraint of the magnetization orientation against the interaction fields and the external fields, it is natural to normalize the effective magnetic fields by crystalline anisotropy field $H_k = 2K/M$ where M is the magnetization magnitude of each grain. The normalized effective field becomes

$$\mathbf{h}_i = \mathbf{H}_i/H_k$$

$$= (\mathbf{k}_i \cdot \mathbf{m}_i)\mathbf{k}_i + h_m \sum_{j=1}^{N} \mathbf{D}_{ij} \cdot \mathbf{m}_j + h_e \sum_{n.n.} \mathbf{m}_j + \mathbf{h}_a \tag{2.9}$$

where the last term $\mathbf{h}_a = \mathbf{H}_{app}/H_k$ is the normalized external applied field. In the Eq. 2.9,

$$h_m = \frac{M}{H_k} \tag{2.10}$$

is the magnetostatic interaction field constant, measuring the magnetostatic interaction strength relative to the crystalline anisotropy constraint and the coefficient

$$h_e = \frac{A^*}{Ka^2} \tag{2.11}$$

is the intergranular exchange coupling constant, measuring the intergranular exchange interaction strength relative to the crystalline anisotropy constraint. a is the center to center distance between adjacent grains. h_e has the same form as the exchange field form used in classic micromagnetic theory (Brown 1963) and should be considered as a first order approximation to the exact intergranular exchange field. It is also important to point out that h_e is inversely proportional to the square of the grain diameter for zero intergranular boundary separation in which case large grain diameter will yield a smaller intergranular exchange coupling constant with the same A^*.

2.3. Dynamic Equation for Magnetization Orientations

The magnetization direction of a magnetic grain in the array follows the Landau-Lifshitz gyromagnetic equation of motion. The field each grain experiences is the effective magnetic field defined in Eq. 2.8. For an array with total N grains, there are

N vectorial first order differential equations coupled through the magnetostatic interaction field and the intergranular exchange field. The time integration of coupled equations:

$$\frac{d\mathbf{M}_i}{dt} = \gamma\mathbf{M}_i \times \mathbf{H}_i - \frac{\lambda}{M}\mathbf{M}_i \times (\mathbf{M}_i \times \mathbf{H}_i), \quad i = 1, 2,..., N \qquad (2.12)$$

gives magnetization configurations during a magnetization process. For simplicity, we can write this set of equations in a reduced form with introducing a reduced time $\tau = t\gamma H_k$ and a reduced damping constant $\alpha = \lambda/\gamma$:

$$\frac{d\mathbf{m}_i}{dt} = \mathbf{m}_i \times \mathbf{h}_i - \alpha\mathbf{m}_i \times (\mathbf{m}_i \times \mathbf{h}_i), \quad i = 1, 2,..., N \qquad (2.13)$$

In some other similar models introduced previously, the Gilbert form is utilized instead (Victora 1987a and 1987b, Mansuripur and McDaniel 1988). It is easy to see that the Gilbert equation and Landau-Lifshitz equation are mathematically identical. The difference is that in the Gilbert equation, the gyromagnetic motion correlates with the damping motion of the magnetization direction and the damping rate becomes limited (Hass and Callen 1963).

Equation 2.13 is a dissipative equation and the energy dissipation rate is proportional to α. In a static external field, the stable static solutions of Eq. 2.13 are local energy minima. The main essence of this modeling study is to find the path, or paths, reaching some of these energy minima, i.e. to identify the physical transient magnetization processes from an initial static state to a new static state, as the external field varies. It has been found that for an infinitely slowly changing external field, such as simulations of hysteresis loop, calculation results are relatively insensitive to the change of the reduced damping constant α (Victora 1987b, Zhu and Bertram 1988a). However, if the applied field changes rapidly, the reduced damping constant could be important to the outcome results. For all the calculation results presented here, the reduced damping constant was chosen in the range of $\alpha = 0.1$ to $\alpha = 1$.

3. Magnetization Processes and Hysteresis Properties

In this section, the fundamental magnetization process and resulting hysteresis properties of the films are discussed. Most of the longitudinal thin film media are planar isotropic: The macroscopic magnetic properties are isotropic in the film plane. In this review, the discussion will be limited for two types of planar isotropic films: The crystalline easy axes of the grains are randomly oriented either in three dimensions (3D-random) or in the film plane (2D-random). The calculation array contains 64×64 grains, corresponding to an approximate $2\mu m \times 1.7\mu m$ area with $D = 300\text{Å}$ assumed.

Figure 3.1 shows three calculated hysteresis loops for a 3D-random film. With $h_m = 0$ and $h_e = 0$, the hysteresis loop (long dashed curve) is the result of assembly of noninteracting Stoner-Wohlfarth particles with the coercivity around $0.48H_k$. Introduction of only the magnetostatic interaction with $h_m = 0.3$ and $h_e = 0$, yields an increase of the saturation remanence and a reduction of coercivity, as the short dashed curve shows. The hysteresis loop becomes "squared-up". Intergranular exchange coupling yields a further increase of the remanence and a further reduction of the coercivity, as the solid curve shows. It also causes much higher coercive squareness S^*. The variations between theses three curves represent essential effects of the two kinds of magnetic interactions to the magnetic hysteresis properties.

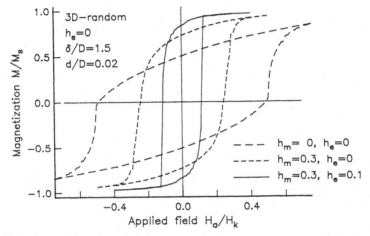

Figure 3.1 Calculated hysteresis loops for 3D-random planar isotropic films. The long dashed curve represents result with no interactions: $h_m = 0$ and $h_e = 0$; the short dashed curve represents the result with only the magnetostatic interaction: $h_m = 0.3$ and $h_e = 0$; the solid curve represents the result with both types of interactions: $h_m = 0.3$ and $h_e = 0.1$.

3.1. Ripple Patterns and Magnetization Remanence

In planar isotropic thin films, even though the orientations of crystalline anisotropy easy axes are random, at the saturation remanent state the magnetizations of grains usually deviate from the local anisotropy easy axes due to the interactions in the film. The magnetization of the film forms a ripple-cluster pattern. A typical magnetization configuration at the saturation remanent state is shown in Fig. 3.2, simulated with $h_m = 0.4$ and $h_e = 0.1$. The local coherence of the magnetization orientation is apparent: a cluster of grains has common magnetization orientation and large changes of the magnetization direction occur between adjacent clusters. Along the initial saturation direction, cluster magnetization directions alternate signs in the transverse direction with apparent quasi-periodicity: The magnetization "ripples" in the film. This cluster-ripple structure characterizes the saturation remanent state in the planar isotropic films

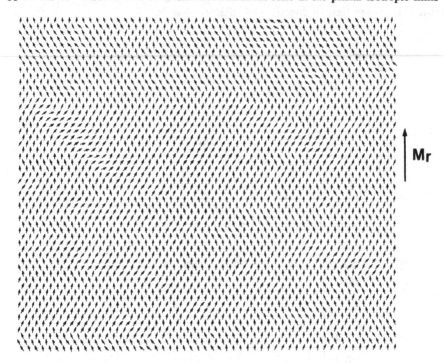

Figure 3.2, A typical magnetization pattern at the saturation remanent state for a 3D-random planar isotropic film. $h_m = 0.4$ and $h_e = 0.1$ were used in the calculation. (Zhu and Bertram 1991c, © IEEE 1991).

and produces a "feather-like" Fresnel mode Lorentz electron microscope image(Chen 1981, Zhu 1989). The ripple pattern is the net result of both the randomly oriented crystalline easy axes and magnetostatic interactions among the grains. The intergranular exchange coupling results in an increase of the cluster size as well as an increase of the wavelengths in the quasi-periodicity along the initial saturation direction (Zhu 1989, Beardsley and Speriosu 1990, Duan et al. 1990). The resulting saturation squareness as a function of h_m and h_e is plotted in Fig. 3.3 and Fig. 3.4, respectively. An initial increase of h_m results in an increase of the squareness for both 2D-random and 3D-random cases. However, if the magnetostatic interaction strength is much stronger than the local crystalline anisotropy field, large angles occur between the

190

magnetizations of adjacent clusters, resulting in local closure of magnetization flux which could lead to formations of magnetization vortices at the saturation remanent state. In comparison, increasing the intergranular exchange coupling only yields a monotonic increase of the saturation squareness along with a decrease of the angles between the magnetization directions of adjacent clusters.

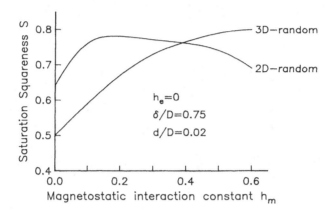

Figure 3.3, Squareness vs. magnetostatic interaction field constant h_m for 3D-random case.

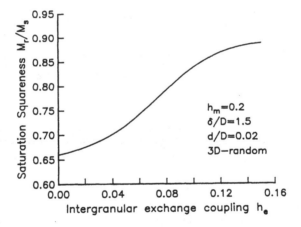

Figure 3.4, Squareness vs. intergranular exchange coupling constant h_e for 3D-random case.

191

3.2. Magnetization Reversal Process

Through a series of simulation studies (Zhu and Bertram 1988a, 1991b, and 1991c, Zhu 1989), it is found that a magnetization reversal process in planar isotropic longitudinal thin films can be characterized by formation of magnetization vortices at the beginning of the reversal, followed by vortex motion and expansion of reversed regions, and ending with annihilations of the magnetization vortices.

(1) *Vortex formation*. From the simulation studies, it has been found that magnetization vortex is the most elementary structure during a magnetization reversal. A reversed region nucleates by formation of a magnetization vortex. Fig. 3.5 shows a typical vortex formation process, simulated with $h_m = 0.4$ and $h_e = 0.1$. The picture 1 represents the magnetization configuration at the saturation remanent state with a typical ripple pattern. As the external field increases in the reverse direction, the ripple pattern develops to a "⊂" structure (picture 3) and this process is usually reversible. With further increasing the applied field, the "⊂" structure becomes unstable and the magnetization at the open-end of the "⊂" undergoes transient irreversible rotations, resulting in the formation of the vortex. The magnetization rotation at the open-end of the "⊂" is essentially a fanning mode reversal (Bertram and Mallinson 1969). Magnetostatic interaction is the intrinsic driving force for the vortex formation which leads to local closure of magnetization flux. The vortex formation can be considered as a

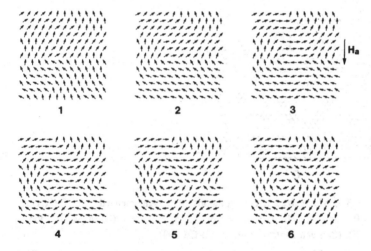

Figure 3.5, A typical vortex formation process, calculated with $h_m = 0.4$ and $h_e = 0.1$. $\delta/D = 1$ and $d/D = 0.02$ were used for the film parameters. Pictures 1,2, and 3 are static states with increasing field values. Picture 3-6 represent a transient process leading to a vortex formation. (Zhu and Bertram 1991c, © IEEE 1991)

192

natural development from the cluster-ripple structure. Figure 3.6 shows three magneti-zation configurations at the simulated "ac" demagnetized state for various interaction parameters. The demagnetized states were simulated by initially setting the magnetiza-tion normal to the film plane followed by relaxation at zero external applied field. As shown in the figure, increasing magnetostatic interaction strength, h_m, results in better defined vortex structure (B) and increasing the intergranular exchange coupling yields significantly larger vortex sizes and much greater adjacent vortices separation (C).

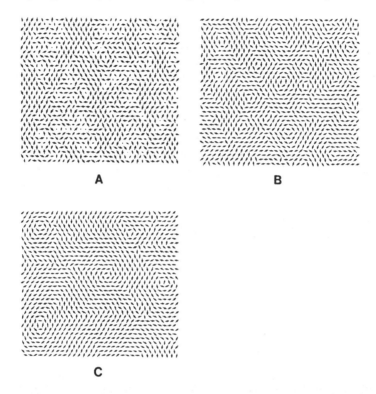

A   B

C

Figure 3.6, Magnetization configurations of a demagnetized state for (A):
$h_m = 0.3$ and $h_e = 0$, (B) $h_m = 0.6$ and $h_e = 0$, and (C) $h_m = 0.3$ and
$h_e = 0.2$. (Zhu and Bertram 1991c, © IEEE 1991).

(2) *Vortex motion.* Magnetization vortex formation results in a nucleation of a reversed region. A reversed area can expand through motion of the magnetization vortex. For zero or weak intergranular exchange coupling, a vortex motion is characterized by a two step process: elongation of vortex center followed by contraction of the elongated vortex center to a new position. A typical two-step process is shown in Fig. 3.7,

calculated with $h_m = 0.4$ and $h_e = 0.1$. The elongated vortex center often is a static state (Picture 3) from which the transient contraction process moves the vortex center

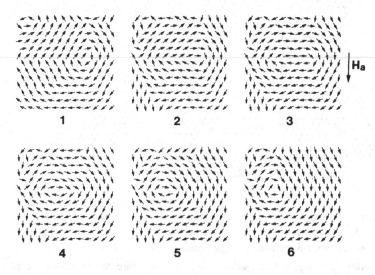

Figure 3.7, A typical vortex motion process calculated with weakly exchange coupling $h_e = 0.1$. $h_m = 0.4$, $\delta/D = 1$, and $d/D = 0.02$ were used in the calculation. Pictures 1-4 illustrate the elongation process of the vortex center and pictures 4-6 illustrate the contraction of the vortex center. (Zhu and Bertram 1991c, © IEEE 1991)

to a new position (Picture 3-6). The vortex motion is mainly in the direction transverse to the reversal field, resulting in expansion of the reversed region. The vortex motion distance in a single movement is virtually determined by the elongation distance. The motion distance varies throughout the film (the calculation array) due to the spatial randomness of the easy axes orientation. With zero intergranular exchange coupling, the motion distance is always very limited, however increasing magnetostatic interaction strength does gradually increase the motion distance. Intergranular exchange coupling significantly enhances vortex motion distance. If the intergranular exchange coupling is strong, the vortex motion process becomes continuous rather than the two-step process and the motion distance in a transient process is often much larger than that in the two-step motion process. Fig. 3.8 shows a vortex motion process calculated with $h_e = 0.2$. The entire motion process from pictures 1 to 6 is transient. Such continuous vortex motion leads to large vortex motion distance and the corresponding expansion of the reversed region yields a wide reverse "domain".

194

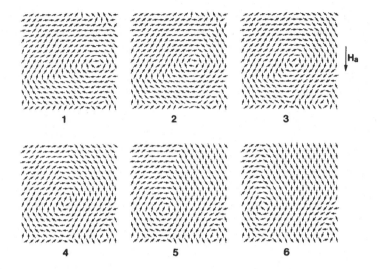

Figure 3.8, A typical vortex motion process for a strongly exchange coupled film, $h_e = 0.15$. $h_m = 0.4$, $\delta/D = 1$, and $d/D = 0.02$ were used for the calculation. Pictures 1-6 represent a single transient motion process.

(3) *Formation of elongated domain*. The ripple structure at the saturation remanent state not only leads to formation of a single vortex but also leads to a collective formation of a series of vortices along the applied reverse field direction and results in an elongated reverse domain. (The term "domain" here is loosely defined as compared with the concept of domain in soft magnetic films.) Figure 3.9 demonstrates a typical formation process of an elongated reverse domain for a film with zero intergranular exchange coupling: $h_m = 0.6$ and $h_e = 0$. At an applied reverse field, the ripple structure develops into a series of alternating "⊂" and "⊃" patterns (picture 1-2). At a larger field value, the "⊂" and "⊃" series evolves to a series of vortices along the field direction, with adjacent vortices having opposite sense of rotation (picture 3). Further increasing the reversal field results in transverse motion of the vortices. Vortices with opposite sense of rotation move in opposite directions and the reversed region in between expands, resulting in the reverse domain elongated along the field direction (picture 4). Each side boundary of the reversed domain contains a series of vortices with same sense of rotation. Since in the films with zero intergranular exchange coupling vortex motion distance is always very limited, the elongated reverse domains are narrow (only slightly larger than vortex diameters) and densely formed narrow elongated domains characterize magnetization pattern around the coercive state. A reversal process calculated with $h_m = 0.3$ and $h_e = 0$ is shown in Figure 3.10, with the gray scale representing the magnetization component in the direction of the external

195

field. Elongation of the reverse domains (dark regions) along the field direction can be

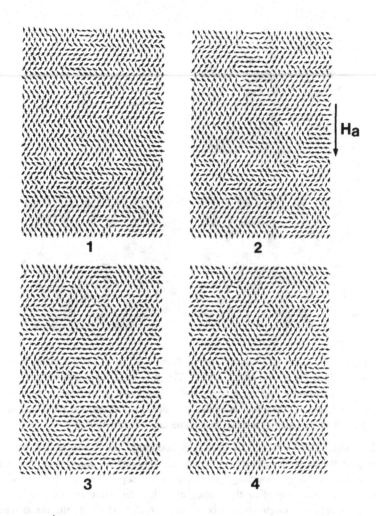

Figure 3.9. A typical formation process of an elongated reverse domain for an nonexchange coupled case, $h_e = 0$ and $h_m = 0.6$. The magnetization configurations 1-4 corresponding to four static states along a major hysteresis loop at increasing external reversal field values. (Zhu and Bertram 1991b)

clearly seen and near the coercive state, the magnetization pattern shows densely distributed elongated domains, characteristic for nonexchange coupled films.

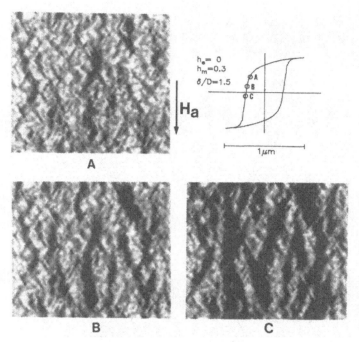

Figure 3.10. Magnetization patterns of three static states along a major hysteresis loop for $h_e = 0$ and $h_m = 0.3$. The gray scale represents magnetization component along the external field direction with *full bright* for the magnetization in the initial saturation direction and *full dark* for the direction parallel to the external reversal field. (Zhu and Bertram 1991b)

For films with intergranular exchange coupling, the essential features in the formation process of an elongated reverse domain remains the same as in the case of zero intergranular exchange coupling. However, the separation of the vortices in the domain boundaries significantly increases with an increase of the intergranular exchange coupling strength, which is a natural consequence of increased wavelengths in the ripple structure due to the exchange coupling. In between the adjacent vortices along each boundary of the domain, magnetization cross-tie structure becomes rather evident, similar to the crosstie structure observed in soft films but with much smaller scale (Moon 1959). In films with relatively strong intergranular exchange coupling ($h_e > 0.15$), vortex motion usually follows the formation of an elongated domain without increasing the external reversal field and the motion of the vortices in a

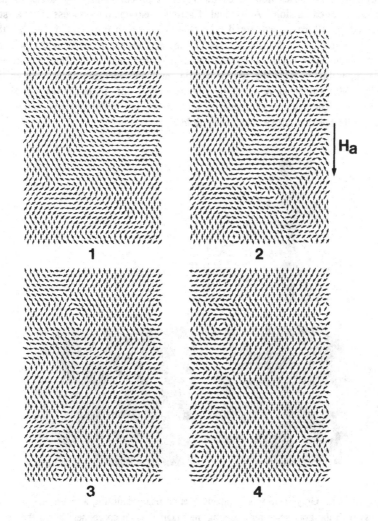

Figure 3.11. A typical formation process of an elongated reverse domain for a strongly exchange coupled case, $h_e = 0.2$ and $h_m = 0.3$. Pictures 1-4 represent a single transient process during a magnetization reversal with the external reversal field remaining constant. (Zhu and Bertram 1991b)

domain boundary becomes rather collective, leading to significant expansion of a reversed domain. The domain expansion through the collective motion of the vortices in the domain boundaries dominates the reversal process and large areas of the film reverse by the expansion. A typical formation-expansion process for a strongly exchange coupled film (h_e = 0.2 and h_m = 0.3) is shown in Fig. 3.11. However, the collective vortex motion in the domain boundaries are not as coherent as the wall motion in soft magnetic films since the spatial variation of the local crystalline anisotropy field is still comparable to both the magnetostatic and the intergranular exchange interaction fields. Magnetization reversal by domain expansion in exchange coupled films results in high (nearly unity) coercive squareness and relatively low coercivity (as compared with the crystalline anisotropy field H_k).

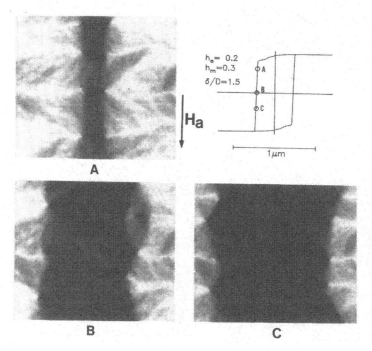

Figure 3.12. Gray scale plot of magnetization patterns during a reversal process with the gray scale representing the magnetization component along the external reversal field direction. The expansion of the elongated reverse domain without an increase of the external reverse field yields a near unity coercive squareness. (Zhu and Bertram 1991b)

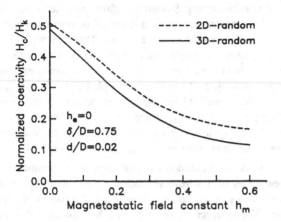

Figure 3.13. Coercivity vs. magnetostatic interaction field constant h_m for nonexchange coupled 2D-random (dashed curve) and 3D-random (solid curve) films ($h_e = 0$).

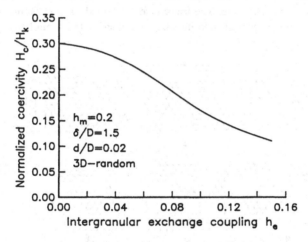

Figure 3.14. Coercivity vs. intergranular exchange coupling constant h_e for 3D-random case. Similar behavior is also obtained for the 2D-random case.

200

The collective magnetization process resulting from magnetostatic and intergranular exchange interactions significantly changes the hysteresis properties in comparison with the behavior of non-interacting Stoner-Wohlfarth particles. An increase of the magnetostatic interaction strength results in a decrease of film coercivity as shown in Fig. 3.13 for both 2D-random and 3D-random cases. The formation of magnetization vortices driven by magnetostatic interactions causes local irreversible magnetization reversals occurring at smaller external reversal field as compared with the case without magnetostatic interaction. In exchange coupled films, magnetization reversal in large areas in the film is realized through the expansion of reversed domains by vortex motion. Coercivity significantly decreases with increasing the intergranular exchange coupling. Along with the coercivity reduction, the coercive squareness significantly increases. The calculation results are shown in Fig. 3.14.

(4) *Vortex annihilation*. A magnetization reversal ends with annihilations of magnetization vortices. Contraction of a unreversed domain leads to merging of the domain boundaries on opposite sides of the domain, resulting in formation of vortex-vortex, or vortex-crosstie pairs. The unreversed domain vanishes through the annihilation of these pair structures. Figure 3.15 presents a typical annihilation process of a vortex-crosstie pair with $h_e = 0.1$ and $h_m = 0.4$. Picture 1 shows the vortex-crosstie pair prior to the annihilation process with the vortex and the crosstie aligned transverse to the external reversal field. The annihilation process begins as the magnetization in between the vortex and the crosstie rotates into the transverse direction (pictures 2, 3, and 4), yielding a "⊂" structure. The transversely oriented magnetizations in the "⊂" structure rotate into the reversal field direction, completing the annihilation process.

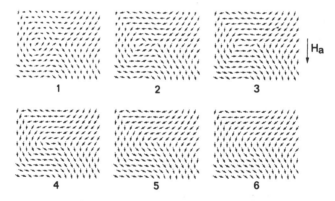

Figure 3.15. A typical annihilation of a vortex-crosstie pair in a weakly exchange coupled film: $h_e = 0.1$ (3D-random). Parameters used in the calculation are $h_m = 0.4$, $\delta/D = 1$, and $d/D = 0.02$.

Two vortices with opposite sense of rotation (since they always come from oppo-site sides of a unreversed domain) often form a vortex-vortex pair followed by a vortex-vortex pair annihilation. A representative annihilation process is shown in Fig. 3.16. The annihilation starts as the vortices align themselves in the applied field direc-tion and the vortices begin to open, forming a pair of "⊃" and "⊂" structures aligned along the field direction (pictures 1-4). The process completes as the transversely oriented magnetizations rotate towards the applied field direction (pictures 5 and 6).

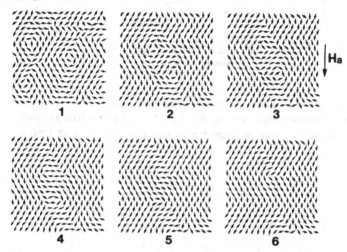

Figure 3.16. A typical annihilation process of a vortex-vortex pair in a weakly exchange coupled film: $h_e = 0.1$. Other parameters used in the cal-culation are $h_m = 0.4$, $\delta/D = 1$, and $d/D = 0.02$ with 3D-random crystal-line easy axis orientations. Picture 1 represents the static state prior to the annihilation. Pictures 2-6 represent the transient annihilation process. (Zhu and Bertram 1991b)

3.3. Hysteresis Properties and Film Microstructures

The change of microstructural parameters, such as intergranular boundary separa-tion and film thickness, alters the magnetostatic interaction strength, thereby changing the hysteresis behavior. In general, coercivity decreases with increasing film thickness, as shown in Fig. 3.17. The coercivity increases with an increase of the intergranular boundary separation, provided that the grain height (i.e. the film thickness) to grain diameter ratio δ/D is less than one, as shown in Fig. 3.18. Increase intergranular boun-dary separation yields an increase of the coercivity due to an effective reduction of magnetostatic interaction strength.

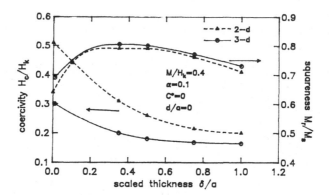

Figure 3.17. Calculated coercivity vs. normalized film thickness δ/D for zero intergranular exchange coupling. $h_m = 0.4$ and $d/D = 0.02$ are used in the calculation. Exchange coupled films also show similar trend. (Zhu and Bertram 1988a)

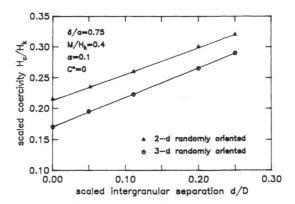

Figure 3.18. Coercivity as a function of scaled intergranular boundary separation d/D for 2D-random and 3D-random cases with zero intergranular exchange coupling. The grain height-to-diameter ratio is kept at a constant: $\delta/D = 0.75$. (Zhu and Bertram 1988a)

For $\delta/D > 1$, the situation becomes slightly more complicated, but interesting. Figure 3.19 shows calculated coercivity as a function of inverse grain separation,

$a = D + d$, with constant grain height-to-diameter ratio $\delta/D = 1.5$. In this case, the shape anisotropy of each individual grain becomes normal to the film plane and well separated grains have very small in-plane coercivity. Reduction of grain boundary separation results in significant increase of film coercivity. The magnetostatic interactions among the grains produces an in-plane anisotropy field. At $\delta/(D+d) = 1.4$, coercivity reaches a maximum. As the grain boundary separation decreases further to near contact, the magnetostatic interaction strength becomes so strong that the coercivity decreases.

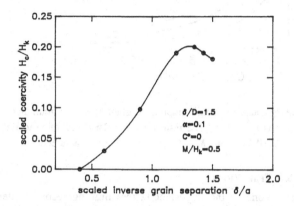

Figure 3.19. Coercivity vs. scaled inverse intergranular separation d for a constant greater-than-unity grain height-to-diameter ratio: $\delta/D = 1.5$. The crystalline easy axes orientation is 3D-random and intergranular exchange coupling constant is zero. (Zhu and Bertram 1988a)

For the films with zero intergranular exchange coupling, hysteresis properties only depend on the ratio of the film thickness and grain diameter δ/D and the ratio of intergranular boundary separation d/D, independent of the actual value of grain diameter D, provided that D is not so small that superparamagnetic behavior occurs and also not so large that the grain becomes multidomain. However, if the intergranular exchange coupling is non-zero, the hysteresis property would depend on actual value of grain diameter D since to the first order approximation, the intergranular exchange coupling constant h_e is inversely proportional to the square of grain diameter D (as discussed in Sec. 2) and changing h_e significantly changes hysteresis properties. Figure 3.20 plots h_e versus D for various values of A^*. Because coercivity increases with decreasing the intergranular exchange coupling constant h_e, increasing grain diameter D yields an increase of film coercivity. Such coercivity behavior is often observed experimentally (Arnoldussen 1986, Johnson et al. 1990).

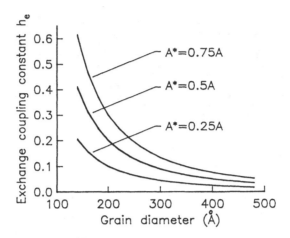

Figure 3.20. Intergranular exchange coupling constant h_e versus grain diameter D with various A^* values. The results are calculated according to Eq. 2.11 with $A = 1.6 \times 10^{-6} erg/cm$, $K = 10^6 erg/cm^3$. $d = 0$ is assumed.

4. Noise of Recorded Transitions

Transition noise is one of the important factors limiting recording densities. In longitudinal thin films, noise is mainly located in the magnetization transition centers, yielding a linear increase of the total noise power with an initial increase of recording density. Furthermore at high recording densities, noise at transitions can be significantly enhanced. The micromagnetic simulation model has been utilized to study the transition noise behavior. In this section, the studies of noise for isolated and interacting transitions are reviewed. The discussion will be limited to planar isotropic longitudinal thin films.

Medium noise arises from spatial magnetization fluctuations which cause variations in the recorded transitions. In order to study the noise behavior, it is necessary to study a large number of transitions and their statistical behavior. For all the statistical results presented in this section, an assembly of $n = 80$ isolated transitions or transition pairs were simulated for each case studied. Each transition, or each transition pair, in an assembly is simulated with a different distribution of randomly oriented grain crystalline easy axis to mimic recording at different locations of the medium. The spatial variations of other intrinsic parameters may also be important but they will not be discussed in this section.

Since a reproducing head averages magnetization flux across the trackwidth, it is proper to study the fluctuations of the track-width averaged transition profiles:

$$M(x) = \frac{1}{W} \int_{W/2}^{W/2} M(x,z)dz \qquad (4.1)$$

where x-axis is along the recording track direction and z-axis is in the cross-track direction. $M(x,z)$ is the simulated magnetization pattern and $M(x)$ is the trackwidth averaged transition profile. For the transition profile $M(x)$, only the magnetization component along the track direction (x component) is analyzed. The simulation array contains 64×128 grains, an approximate 3.6μm long and 2μm wide recording track with $D = 300$Å The mean transition profile of the transition ensemble is:

$$\bar{M}(x) = \frac{1}{n}\sum_{i=1}^{n} M_i(x) , \qquad (4.2)$$

and the variance of the transition profile ensemble is

$$\overline{\Delta M^2}(x) = \frac{1}{(n-1)}\sum_{i=1}^{n} [M_i - \bar{M}(x)]^2 \qquad (4.3)$$

The ensemble variance of the transition profiles represents the noise. An total variance, corresponding to the measured total noise power, is defined as

$$NP = \frac{1}{M_r^2 B} \int_{x_c-B/2}^{x_c+B/2} \overline{\Delta M^2(x)}dx \qquad (4.4)$$

where x_c is the mean transition center position. The integrated variance of the transition profiles resembles the total noise power one measures on the spectrum analyzer since

$$\lim_{L\to\infty} \frac{1}{L} \int_{-L/2}^{+L/2} |\Delta M(x)|^2 dx = \frac{1}{2\pi} \lim_{L\to\infty} \frac{1}{L} \int_{-\infty}^{+\infty} |\Delta \hat{M}(k)|^2 dk. \qquad (4.5)$$

4.1. Noise of Isolated Transitions

The micromagnetic model has been utilized to study transition noise behavior of isolated transitions (Zhu and Bertram 1988b and 1990, Bertram and Zhu 1991). One of important findings in the simulation studies is that the intergranular exchange coupling results in large transition noise. It is also found that transition noise behavior shows no distinguish difference between the films with crystalline easy axes randomly oriented in the film plane and the films with the easy axes randomly oriented in three dimensions.

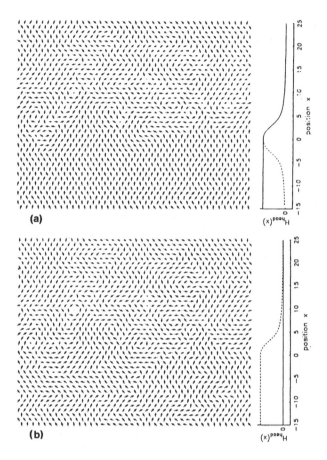

Figure 4.1, In-plane projection of the magnetization transition configuration: (a) the static state with the head field applied and (b) the static state after the head field off. Vectorial Karlqvist head field was used and the crystalline easy axes of the grains are randomly oriented in three dimensions. (Zhu and Bertram 1988b, © IEEE 1988)

In the earlier studies, a transition was simulated by relaxation from an initially created perfect step head-on transition (Zhu and Bertram 1988b). However, it has been found that the transition noise does depend on the gradient of the recording field (Tanaka et al. 1982, Johnson et al. 1991). Transition noise tends to increases as field gradient decreases, partially because the resulting transition becomes broader. Therefore, Karlqvist head field function was utilized with typical gap-length and flying heights in latter studies (Zhu and Bertram 1990, Bertram and Zhu 1991). To simulate the motion of the medium, the head field values at the gap center of a recording head are extended to the leading edge of the head. The simulation array is initially saturated in the opposite direction of the recording field and relaxed to the saturation remanent state. The head field is applied with zero rise time. The head field remains unchanged until the magnetization reaches a static state. A final transition configuration is obtained by setting the head field to zero and letting magnetizations relax to the static configuration. Figure 4.1 shows the magnetization configurations at the static state with head field on (a) and the final configuration with the head field off (b).

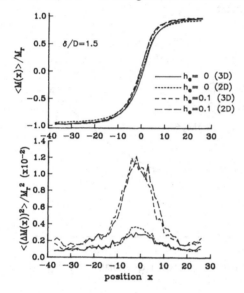

Figure 4.2. Ensemble mean of the track-width-averaged transition profile and ensemble variance with ($h_e = 0.1$) and without exchange coupling for 2D-random and 3D-random cases. $h_m = 0.2$ was used in the calculations for the exchange coupled cases and $h_m = 0.3$ was used for the nonexchange coupled cases so that the mean transition lengths are similar for both cases (Bertram and Zhu 1991, © IEEE 1991).

Figure 4.2 shows the ensemble mean and the ensemble variance of simulated transition profiles for films with different intergranular exchange coupling constant: $h_e = 0$ and $h_e = 0.1$, as well as different easy axes orientations: The crystalline easy axes are randomly oriented either in three dimensions (3D-random) or in the film plane (2D-random). Different values of h_m were used for films with different exchange coupling constant so that the ratio M_r/H_c becomes similar in the two cases.

As shown in the figure, the normalized mean transition profiles for the four cases are rather similar. However, the variance of the transition profile ensembles shows a large difference between the films with and without the intergranular exchange coupling. For films with $h_e = 0$, the variance exhibits a small increase at the center of transition for both 2D-random and 3D-random cases. However, for films with $h_e = 0.1$, the variance at the transition center becomes significantly higher than the non-exchange coupled cases. The peak values of the variance for the exchange coupled cases are over four times larger than that for the non-exchange coupled cases. The peak values are very similar between the 2D-random and 3D-random cases with the same intergranular exchange coupling constant.

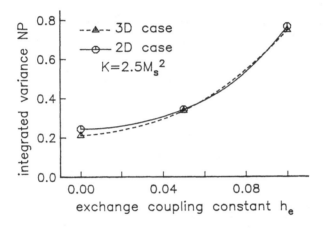

Figure 4.3. Total variance over the transition region vs. intergranular exchange coupling strength with $h_m = 0.2$ for both 2D-random and 3D-random cases.

The intergranular exchange coupling results in significant increases of transition noise. In Fig. 4.3, the total variance (bit length B was chosen to be the distance from $-0.8M_r$ to $0.8M_r$) is plotted versus the exchange coupling constant. The total variance increases with an increase of the exchange coupling constant and have similar values in the both 2D-random and 3D-random cases. The large transition noise is caused by large scale irregular transition boundaries resulting from the intergranular exchange coupling and the two types of random easy axis orientations result in little difference in the transition noise properties. Figure 4.4 shows gray scale plot of typical transition patterns for the exchange coupled case (a) and the case with zero intergranular exchange coupling (b). The gray scale represents the magnetization component in the recording field direction. The *full bright* represents the magnetization direction along the recording field and *full dark* represents the opposite direction. For the exchange coupled case, the transition boundaries exhibit large scale irregular zig-zag structures. These large scale irregular zig-zag patterns result in large fluctuations in the trackwidth averaged transition profiles, thereby large transition noise. For zero exchange coupling, the transition boundary is characterized by narrow fingers thereby much smaller fluctuations in the trackwidth averaged transition profiles. In Fig. 4.5, vector plots of two typical magnetization transition configurations for the exchange coupled (a) and nonexchange coupled (b) cases are shown. For the exchange coupled case, magnetization vortices are formed at the center of the transition and rather irregularly distributed along the center of the transition. Away from transition center, magnetization exhibits a ripple pattern. Near the transition center, "⊂" and "⊃" structures can be observed due to the demagnetization field. For the non-exchange coupled case, the vortices are not as well defined as those in the exchange coupled case and the size of the vortices are much smaller.

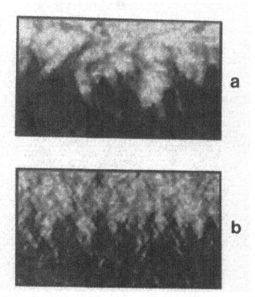

Figure 4.4. Typical magnetization patterns of simulated recorded isolated transition for the exchange coupled case with $h_e = 0.1$ and $h_m = 0.2$ (a) and the nonexchange coupled case: $h_e = 0$ and $h_m = 0.3$ (b). The gray scale represents the magnetization component along the recording direction and the track-width is 1.98 μm, The crystalline easy axes orientation is 3D-random.

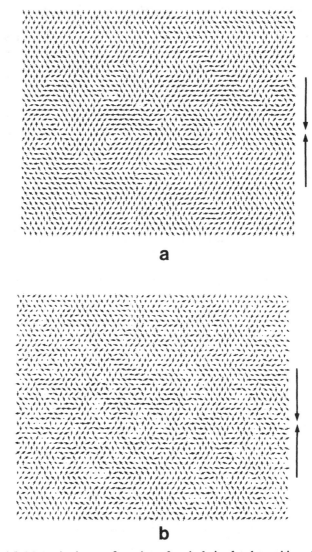

a

b

Figure 4.5. Magnetization configuration of typical simulated transitions for the exchange coupled case: $h_e = 0.1$ and $h_m = 0.2$ (a) and the zero exchange coupling case: $h_e = 0$ and $h_m = 0.3$ (b). The crystalline easy axes orientation is 3D-random.

4.2. Noise of Interacting Transitions at High Recording Densities

For most longitudinal film media, the integrated noise power increases more rapidly at high recording densities than the initial linear increase. This is often referred to as a supralinear increase and an indication that the amount of noise in the transitions increases at high recording densities. The experimental study on *FeCoCr* films by Arnoldussen and Tong using Lorentz electron microscopy to image recorded bits suggested that the percolation of adjacent transition boundaries at high recording densities could enhance the noise (Arnoldussen and Tong 1986). Madrid and Wood pointed out that the presence of the demagnetization field during the recording process yields a negative correlation between the adjacent transitions at high recording density but solely the negative correlation between the transition could not yield the supralinear increase of the total noise (Madrid and Wood 1986). A theoretical analysis based on Williams-Comstock model was suggested by Barany and Bertram (Barany and Bertram 1987). Their calculation was able to predict the initial departure of the noise power from the initial linear increase by assuming that the variance of a transition position is correlated with the variance of the previous transition's position by the

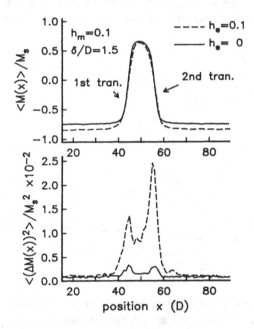

Figure 4.6. Ensemble mean and ensemble variance of simulated di-bit transition profiles for $h_e = 0.1$ (dashed curve) and $h_e = 0$ (solid curve). The interbit interval is $B/D \approx 11$.

demagnetizing field. However, the analysis is only one dimensional and does not taken into account of the two-dimensional feature of the irregular transition boundaries, as shown by Arnoldussen and Tong's experimental study.

Recently, Zhu studied di-bit transition pairs, utilizing the micromagnetic model (Zhu 1991). The study not only was able to predict the supralinear increase of the transition noise at high densities, but also provided underlying physics for the noise enhancement and explanation for different noise behavior for various media. By varying the inter-transition interval, B, in di-bit transition pairs, the effects of interactions between the two transitions in di-bit transition pairs as well as the percolation of transition boundaries were studied. In the study, the magnetization pattern of a di-bit transition pair is simulated by applying a Karlqvist head field (including both longitudinal and vertical components) with a gap length $g_h = 0.24\mu m$, a fly height $d_h = 0.037\mu m$ on a previously saturated state. The motion of the medium is stepwise with step size equal to the grain diameter D. Each motion step is taken only after the magnetization configuration reaches a static state. After the medium is moved by a bit interval, B, the sign of the head field is reversed to generate a second transition. The final magnetization pattern is obtained by setting the head field to zero. Like the noise study for isolated transitions discussed earlier in this section, the only intrinsic

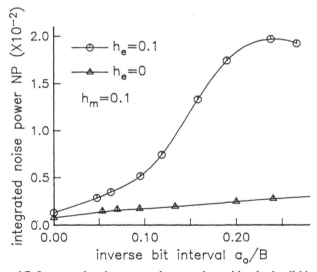

Figure 4.7. Intergrated variance over the second transition in the di-bit transition pairs vs. normalized inverse bit interval. For the exchange coupled case, the supralinear increase of the variance at small bit intervals is evident. (Zhu 1991, © IEEE 1991).

randomness is that the crystalline easy axes of the grains in the calculation array are randomly oriented in the film plane (2D-random).

Films with and without the intergranular exchange coupling were studied. Figure 4.6 shows the mean and the variance of the transition profile ensemble for an exchange coupled case, $h_e = 0.1$, and a nonexchange coupled case, $h_e = 0$. Both cases were calculated with a common $h_m = 0.1$. For the exchange coupled case, the variance exhibits large values at the centers of both the first and second transitions. The variance peak value at the first transition is virtually the same as an isolated transition whereas the value at the second transition is significantly greater. At this bit interval B, the presence of first transition results in higher transition noise at second transition. For the nonexchange coupled case, the variance at transition centers is much smaller than that for the exchange coupled case. Little increase of the variance at the second transition is observed. To study the noise enhancement in the second transition of the di-bits, the total variance calculated only over the second transition (the integral limit is from $-B/2$ to $B/2$ centered at the second transition) is studied. In Fig. 4.7, the total variance of the second transition is plotted as a function of inverse bit interval. a_o in the normalization is the mean transition length parameter for transitions far apart. For the exchange coupled case ($a_o \approx 0.06 \ \mu m$), the total variance increases linearly at large

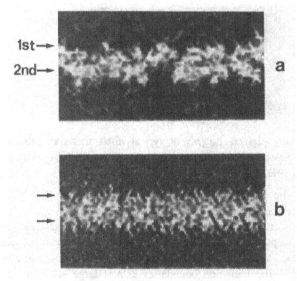

Figure 4.8. Magnetization patterns of typical di-bit transition pairs at $B \approx 6a_o$ for $h_e = 0.1$ (a) and the $h_e = 0$ (b). The gray scale represents magnetization component along the recording direction. The track-width shown here is 1.98 μm wide. (Zhu 1991, © IEEE 1991).

214

bit intervals. When the two transitions in the di-bit are far apart, the variance integrated over the transition center remains the same for both transitions. Since the total variance is defined as the variance integrated over the transition divided by the bit interval, thereby a linear increase of the total variance versus the inverse bit interval. At small bit intervals, i.e. the two transitions in the transition pair are closely placed, the total variance exhibits a significant supralinear increase due to the increase of the variance at the second transition. However, for the case $h_e = 0$ ($a_o \approx 0.05 \ \mu m$), no apparent supralinear increase is shown.

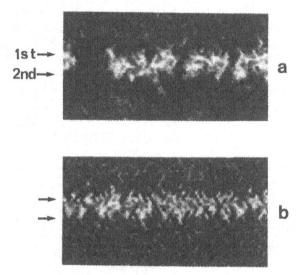

Figure 4.9. Magnetization patterns of typical di-bit transition pairs at $B \approx 5a_o$ for the exchange coupled case (a) and the non-exchange coupled case (b). (Zhu 1991, © IEEE 1991).

To understand the nonlinear increase of the transition noise in the exchange coupled case and the different behavior in the nonexchange coupled case, it is very helpful to study the magnetization patterns of the simulated di-bit transition pairs. Figure 4.8(a) shows the magnetization pattern of a typical di-bit transition pair for the exchange coupled case at $B \approx 6a_o$ where the supralinear increase of the total variance is already significant. The transition boundaries exhibit large scale irregular zig-zag structure. The magnetostatic interaction fields resulting from such transition boundary fluctuate greatly across the recording track. At this bit interval, the fluctuating field significantly modifies the head field during the writing of the second transition and a much noiser second transition results. In Fig. 4.8 (b), the magnetization pattern for the nonexchange coupled case at same bit interval is shown. In this case, the transition boundaries consist of narrow finger structures. The narrow finger transition boundary yields not only very small fluctuations in the track width averaged transition profiles, but also a magnetostatic interaction field with very small fluctuations where a later adjacent transition is to be written, thereby little noise enhancement.

The study also shows that for the exchange coupled case, at the bit interval where the integrated variance of the second transition reaches a maximum, the transition boundaries have just become percolated. Two typical di-bit magnetization patterns for the exchange coupled case (a) and the nonexchange coupled case (b) at this bit interval are shown in Fig. 4.9 The percolated transition boundaries for the exchange coupled case shows the large size of irregular island domains while the percolation "channels" for the nonexchange coupled case are narrow and much denser. In the exchange coupled case, percolation of the transition boundaries yields an increase of the noise at first transition.

In order to characterize the noise in the transition profile, the transition profile of each simulated di-bit is fitted to a pair of linearly superimposed analytical functions through least square fitting,

$$M_i(x) = M_r \left[f(x_{1,i}, a_{1,i}, C_{1,i}) - f(x_{2,i}, a_{2,i}, C_{2,i}) - 1 \right] \quad i = 1, 2, ..., 80 \qquad (4.6)$$

The analytical function $f(x_{n,i}, a_{n,i}, C_{n,i})$ is a linear combination of error function (Barany and Bertram 1987) and arctangent function with the same slope at the center of the transition:

$$f(x_{n,i}, a_{n,i}, C_{n,i}) = C_{n,i} \cdot erf\left(\frac{x - x_{n,i}}{a_{n,i} \sqrt{\pi}} \right) + (1 - C_{n,i}) \tan^{-1}\left(\frac{x - x_{n,i}}{a_{n,i}} \right) \qquad (4.7)$$

where $x_{n,i}$, $a_{n,i}$, and $C_{n,i}$ are the fitted transition center position, transition length parameter and weighting constant for the first ($n = 1$) or the second ($n = 2$) transition of the ith di-bit in the ensemble, respectively. The standard deviation of the first and second transition positions are plotted in Fig. 4.10. In the exchange coupled case, the fluctuations of both transition position and transition length parameter fluctuation have similar relative magnitude with respect to the transition length parameter for isolated transitions. The fluctuations of both position and transition length for the first transition in the di-bit remain constant until the percolations of the transition boundaries occur. For the nonexchange coupled case, the fluctuations are small and remain virtually independent of interbit interval. The fluctuations of transition position and transition length parameter are correlated since they are due to the irregularity of the transition boundaries. The parameter $C_{1,i}$ and $C_{2,i}$, defining the shape of the transitions, also fluctuate significantly.

216

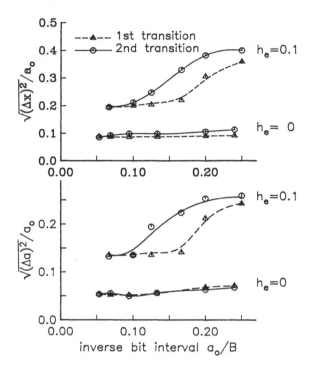

Figure 4.10. Standard deviation of transition center position (a) and transition length parameter (b) of the first transition (dashed curves and triangles) and the second transition (solid curves and circles) for both exchange coupled and nonexchange coupled cases.

5. A Simplified Simulation Model

The simulation studies based on the full dynamic model have been very successful. However, the calculations are extremely computationally intensive, limiting the size of the calculation array. In this section, a much simplified approach is discussed. A cellular automaton model is developed based on the understanding of magnetization process in the thin films gained from the studies utilizing the full dynamic model (Zhu and Bertram 1991). For this simplified cellular automaton model, the computation time is significantly reduced, a much larger size sample can be calculated and reliable statistical behavior of the magnetization domain structure can be obtained.

5.1. Model

In this cellular automaton model, planar isotropic films are represented by closely packed hexagonal cells on a two-dimensional triangular lattice. The magnetization of each cell is allowed only to have two values along the direction of the applied field: $+M_r$ and $-M_r$. Due to this simplification, the calculation results only represent remanent magnetic properties since any rotational reversible process is neglected. A log-normal distribution of local switching fields h_s with dispersion parameter σ is assumed, which is randomly assigned to each cell. In the model, all fields are normalized by the peak field of the distribution. The local switching fields may arise from a distribution in crystalline anisotropy fields. Magnetostatic interactions and intergranular exchange coupling are included, not as in a typical local field model, but in a manner so that collective reversal, such as reversal through formation and expansion

Magnetostatic Interaction   Exchange Interaction

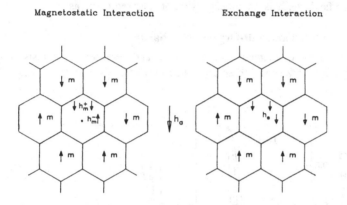

Figure 5.1 Illustration of calculating magnetostatic interaction fields h_m^+ and h_m^- (a) and intergranular exchange fileds (b) for an unreversed cell h_e (the center cell in both figures).

of elongated reverse domains, is incorporated. During a reversal process, both the magnetostatic and the exchange fields on an unreversed cell are assumed to arise only from nearest-neighbor cells that have previously undergone reversal. The interaction field on an unreversed cell from all its reversed nearest-neighbor cells are added to the external applied field. If the total field does not exceed the assigned local switching field for every cell in the array, the applied field is increased until the total field of one cell in the array exceeds its local switching field. Then, the magnetization of this cell is reversed. This process often yields a sequence of magnetization reversals of connected cells due to the changes of the interaction fields resulted from the reversal of the cell. Simultaneous reversals are not allowed; only the magnetization of the cell with maximum difference between the total field and the local switching field reverses its sign. The applied field is not changed until the total field of every cell in the array no longer exceeds the local switching field.

Figure 5.1(a) illustrates the magnetostatic interaction fields for an unreversed cell, the center cell. The field only arise from the nearest-neighbor cells whose magnetization have reversed previously. The field from the reversed nearest-neighbor cells in the direction along the magnetization is in the reverse direction (the direction of the applied field) with magnitude h_m^+ and the field from the reversed nearest-neighbor cells in the directions transverse to the applied field are in the opposite direction (opposite to the applied field) with magnitude h_m^-. This way of incorporating the magnetostatic interaction fields characteristically represents the properties of magnetostatic interaction fields. The intergranular exchange field h_e illustrated in Fig. 5.1(b), is in the reversed direction with same magnitude from all the nearest-neighbor reversed cells. This exchange field yields isotropic expansion of reversed domains.

5.2. Domain Patterns and Self-Organized Behavior

Using this model, magnetization processes and remanent hysteresis properties were studied. Figure 5.2 shows three calculated hysteresis loops with various values of

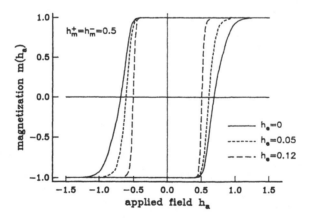

Figure 5.2. Calculated remanent hysteresis loops for various intergranular exchange field value. $h_m^+ = h_m^- = 0.5$ were used for the calculations. The intrinsic switching field distribution is the same for all three loop calculations. (Zhu and Bertram 1991a)

intergranular exchange field $h_e = 0, 0.05, 0.12$ with common values of $h_m^+ = h_m^- = 0.5$. Clearly, increasing the intergranular exchange coupling results in an increase of the remanent coercive squareness S_r^*. For zero intergranular exchange field, reversed domains are narrow and elongated in the applied field direction. Domain widths increase with increasing exchange field, due to the transverse expansion of reversed domains resulting from the exchange field. Figure 5.3 shows domain patterns at partially reversed state $\overline{M} = 0.3M_r$ for $h_e = 0$ and $h_e = 0.12$. These simulated domain patterns are characteristically in good agreement with experimental observations (Nguyen et al. 1990) and their general behavior with the intergranular exchange coupling agrees well with calculated results shown in Sec. 3.

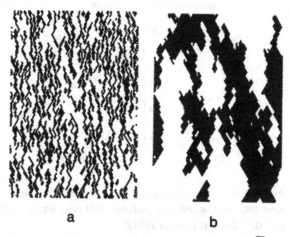

a   **b**

Figure 5.3. Magnetization pattern at partially reversed state: $\overline{M} = 0.3M_r$ for an nonexchange coupled case: $h_e = 0$ (a) and an exchange coupled case: $h_e = 0.12$ (b). $h_m^+ = h_m^- = 0.5$ were used for the calculations. The domains elongate in the external field direction. (Zhu and Bertram 1991a)

During magnetization reversal, an initial reversal of a single cell is often followed by a sequence of magnetization reversals of connected cells, or an avalanche, without increasing the applied field. Magnetostatic interaction fields as well as intergranular exchange fields are the causes for avalanche reversals, resulting in magnetization patterns shown in Fig. 5.3. The magnetostatic interaction fields yield the domain elongation in the field direction and the exchange coupling fields yield the domain expansion. The study found that the frequency of the avalanches as a function of the avalanche sizes near the coercive state follows a power-law:

$$D(S) = AS^{-\alpha} \qquad (5.1)$$

220

where S is the size of the avalanches, the number of cells reversed in a single reversal sequence. Simulation study shows that α decreases with increasing h_e. The intergranular exchange field enhances avalanche sizes and results in wide domains. In Fig. 5.4, the avalanche frequency as a function of avalanche size is plotted for two exchange field values. $\alpha = 1.1$ is obtained for $h_e = 0.12$ and $\alpha = 1.45$ for $h_e = 0.07$. Such power-law behavior of the avalanches indicates that the magnetization patterns around the coercive state is in a self-organized critical state (Bak et al. 1987). The applied reversal field keeps the system marginally stable, whereas the interactions permit reversal propagations on all scales.

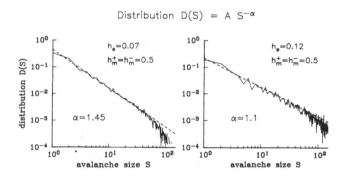

Figure 5.4. An avalanche size distribution during a reversal near the remanent coercive state for two different exchange field constant: $h_e = 0.07$ (a) and $h_e = 0.12$ (b). (Zhu and Bertram 1991a)

5.3. Wohlfarth Relation and Intergranular Exchange Coupling

Recent experimental studies were conducted to investigate the interaction effect in thin-film media by examining the Wohlfarth relation, often referred as the ΔM curve, defined as:

$$\Delta \overline{M}(H_a) = \frac{1}{M_r}\left[2I(H_a) - [M_r + \overline{M}(H_a)]\right] \tag{5.2}$$

where H_a is the external applied field, $I(H_a)$ is the initial remanent magnetization curve which varies from zero (an ac erased state) to M_r and $M(H_a)$ is the remanent hysteresis loop which varies from $-M_r$ to $+M_r$. For an assembly of noninteracting grains, ΔM vanishes at all fields. Because of the interactions among the magnetic grains, reversal processes become very different when starting from different magnetization states. ΔM is one way of measuring the interactive effect during magnetization process and could be a useful tool for the understanding of the complex interactive phenomena in thin film recording media. Mayor et al. measured ΔM curves for a series of *CoCrPt/Cr* film medium with various *Cr* underlayer thickness (Mayo et al. 1991). For thin *Cr* underlayer thicknesses, the ΔM curves have large positive peak. The positive peak value decreases with increasing *Cr* underlayer thickness. For 3000 Å thick *Cr* underlayer, the ΔM curve becomes essentially negative. The cellular automaton model has been applied to calculate the ΔM curves with various intergranular exchange fileds, attempting to correlate the effect of varying the *Cr* underlayer thickness with the change of intergranular exchange field. Figure 5.5 shows calculated results for various intergranular exchange field values. This series of curves show excellent agreement with the measured results on films with various *Cr* underlayer thickness (Mayo et al. 1991), suggesting that in this series of experimental films, the increase of the *Cr* underlayer thickness results in a decrease of the intergranular exchange coupling strength. This prediction agree well with the electron microscopy studies on the physical grain boundary separations caused by thick *Cr* underlayers (Yogi et al. 1988).

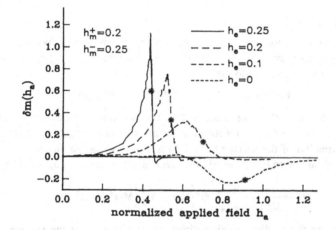

Figure 5.5. Calculated ΔM curves for various intergranular exchange field values. The star symbols marks the corresponding remanent coercive state. (Zhu and Bertram 1991a).

The cellular automaton model is an attempt to develop simplified simulation models which still contain the essential physical characteristics of the magnetization processes. Not only has the model been able to predict measured experimental quantities, but it also has provided new physical insights on the magnetization processes and domain patterns. More sophisticated models, but still remains simple, would further improve our understanding of the magnetization processes in general recording media, including multilayered thin films.

6. Reverse DC Erase Noise

It has been shown experimentally that reverse *dc* erase noise is directly corre-lated with the recorded transition noise in longitudinal thin film recording media (Aoi et al. 1986). This correlation suggests that the noise mechanisms in the two corresponding magnetization processes are related. Understanding the noise properties in reverse *dc* erasure, i.e. at the states along the remanent hysteresis curve, will improve our general understanding of the noise mechanisms as well as the underlying physics governing the magnetization process. Extensive experimental and theoretical studies have shown that medium noise in thin metallic films is dominated by magnetic interactions, especially the intergranular exchange coupling. The intergranular exchange coupling results in large scale irregular transition boundaries, resulting in large noise in recorded transitions. The intergranular exchange coupling also results in large noise at the remanent coercive state (often referred to as the *dc* demagnetized state).

6.1. Analytical Analysis

At a reverse *dc* erased state, i.e. a state along the remanent hysteresis curve, with an average magnetization $\overline{M}$, the read-back voltage power spectrum density under the thin media approximation is

$$|v(k)|^2 = C \times k^2 |H_s(k)|^2 \ e^{-(2d+\delta)} \ M_r^2 \ \delta^2 \ W \ g(k) \tag{6.1}$$

where C is a constant, W is the track-width, δ is the film thickness, d is the fly-height, and M_r is the medium magnetization remanence. In the equation, $H_s(k)$ is the Fourier transformation of the surface head field and $g(k)$ is the Fourier transformation of the along-track autocorrelation function of track-width averaged magnetization:

$$g(x) = \frac{1}{M_r^2} \lim_{L \to \infty} \frac{W}{L} \int_{-L/2}^{+L/2} dx' \left[M(x') - \overline{M} \right] \cdot \left[M(x'+x) - \overline{M} \right] \tag{6.2}$$

where x-axis along the medium motion direction, y-axis is along the surface normal of the erase head, and z-axis is in the cross track direction. $\overline{M}$ is the mean magnetiza-tion averaged over the entire track and the track-width averaged magnetization, $M(x)$, is:

$$M(x) = \frac{1}{W} \int_{-W/2}^{+W/2} dz' \ M(x,z') \tag{6.3}$$

Integrate reduced noise power spectrum $g(k)$ over the entire frequency range:

$$\int_{-\infty}^{+\infty} dk \ g(k) = \frac{1}{M_r^2} \lim_{L \to \infty} \frac{W}{L} \int_{-L/2}^{+L/2} dx \ |M(x) - \overline{M}|^2$$

$$= \frac{1}{M_r^2} \lim_{L \to \infty} \frac{W}{L} \int_{-L/2}^{+L/2} dx \; |\mathbf{M}(x)|^2 - W\left[\frac{\overline{M}}{M_r}\right]^2 \tag{6.4}$$

The first term on the r.h.s. of Eq. (6.4) is:

$$\frac{1}{M_r^2} \lim_{L \to \infty} \frac{W}{L} \int_{-L/2}^{+L/2} dx \; \frac{1}{W^2} \int_{-W/2}^{+W/2} dz' \int_{-W/2}^{+W/2} dz'' \; \mathbf{M}(x,z') \cdot \mathbf{M}(x,z'')$$

$$= \frac{1}{M_r^2} \int_{-W/2}^{+W/2} dz \lim_{L \to \infty} \frac{1}{L} \int_{-L/2}^{+L/2} dx \frac{1}{W} \int_{-W/2}^{+W/2} dz' \mathbf{M}(x,z') \cdot \mathbf{M}(x,z'+z)$$

$$+ \frac{1}{M_r^2} \int_{-W/2}^{+W/2} dz' \lim_{L \to \infty} \frac{1}{L} \int_{-L/2}^{+L/2} dx \; \frac{1}{W} \left[\int_{-W/2-z'}^{-W/2} dz \; \mathbf{M}(x,z') \cdot \mathbf{M}(x,z'+z) \right.$$

$$\left. - \int_{W/2-z'}^{W/2} dz \; \mathbf{M}(x,z') \cdot \mathbf{M}(x,z'+z) \right] \tag{6.5}$$

with change of variable: $z = z''-z'$. The first term on the r.h.s. of Eq. (4) can be written as:

$$\frac{1}{M_r^2} \int_{-W/2}^{+W/2} dz \; f_{c.t.}(z)$$

where $f_{c.t.}(z)$ is the cross-track autocorrelation function:

$$f_{c.t.}(z) = \lim_{L \to \infty} \frac{1}{L} \int_{-L/2}^{+L/2} dx \; \frac{1}{W} \int_{-W/2}^{-W/2} dz' \; \mathbf{M}(x,z') \cdot \mathbf{M}(x,z'+z) \tag{6.6}$$

To approximate the magnetization cross-track correlation function, we choose an analytical function:

$$f_{c.t.}(z) = \frac{s^2(M_s^2 - \overline{M}^2)}{s^2 + (\pi z)^2} + \overline{M}^2 \tag{6.7}$$

which has the correct limits:

$$f_{c.t.}(0) = M_s^2 \quad \text{and} \quad f_{c.t.}(\infty) = \overline{M}^2 \tag{6.8, 6.9}$$

The parameter s in Eq. (6.7) is the magnetization cross-track correlation length. Assuming $W \gg s$, the second term on the r.h.s. of Eq. (6.5) vanishes. Equation (6.4) becomes:

$$\int_{-\infty}^{+\infty} dk \; g(k) = s \times (1 - \overline{m}^2) + \frac{s}{M_r^2}(M_s^2 - M_r^2) \tag{6.10}$$

where $\bar{m} = \bar{M}/M_r$ is the normalized mean magnetization. In general, the cross-track correlation length s should be a function of the magnetization state along the remanent hysteresis curve: $s = s(\bar{m})$. However, for simplicity, let us assume that s is independent of the magnetization states during reverse dc erasure, the second term on the r.h.s. of Eq. (6.10) is a constant and represents the noise magnitude at the saturation remanent state. Define the reduced total noise power:

$$NP_{dc} = \int_{-\infty}^{+\infty} dk f(k, \bar{m} = \bar{m}) - \int_{-\infty}^{+\infty} dk f(k, \bar{m} = 1) \tag{6.11}$$

thereby:

$$NP_{dc} = s \times (1 - \bar{m}^2) \tag{6.12}$$

The noise power during reverse dc erasure is a parabolic function of the normalized mean magnetization and reaches its maximum s at the state $\bar{m} = 0$, provided that s is independent of magnetization state. Although special form of magnetization cross-track correlation function was used in the above derivation, the final result in Eq. 6.12 should not depend on the exact form of the cross-track correlation function as long it has the right limits given by Eq. 6.8 and 6.9. Eq. 6.12 was first derived by Silva and Bertram (Silva and Bertram 1990).

6.2. Computer Modeling Study

In this section, we review a computer modeling study of reverse dc erase noise. The reverse dc erasure process is modeled by simulating the remanent magnetization process. Thus, it is natural to utilize the simple model described in the previous section. The reverse dc erasure process is modeled by simulating the remanent magnetization hysteresis process. The array representing the erasure track contains 5000 cells along the track direction and 100 cells across the track. A spatially uniform external field is applied along the track direction, along x-axis. The medium is initially saturated in the $-x$ direction. Then the positive external field is increased stepwise with very small stepsize. At each field step, the reduced noise power is calculated according to:

$$NP_{dc} = \frac{W}{L} \int_0^L |m(x) - \bar{m}|^2 \, dx \tag{6.13}$$

$$\text{and} \quad m(x) = \frac{1}{M_r W} \int_0^W M(x, z) \, dz \tag{6.14}$$

where W is recording track width (here it is the width of the simulation array), L is the track length (length of the array) and $\bar{m}$ is the normalized magnetization averaged over the entire array. The above expression is exactly the same as Eq. 6.12.

The noise power as a function of the mean magnetization calculated from simulated magnetization patterns (as the ones shown in Fig. 5.2) is plotted (as symbols) in Fig. 6.1 for three different intergranular exchange coupling strengths. The solid curves in the figure represent the fit parabolic function $s(0)\times(1-\overline{m}^2)$. All three curves only follows the function $s(0)\times(1-\overline{m}^2)$ approximately, indicating that the crosstrack correlation length s does depend on the magnetization state along the remanent hysteresis curve: $s = s(\overline{m})$. Increasing the intergranular exchange coupling results in a significant increase of the noise power at the remanent coercive state, i.e. the crosstrack correlation length s significantly increases with increasing the intergranular exchange coupling because of the increase of the domain width, as patterns in Fig. 5.2 demonstrated.

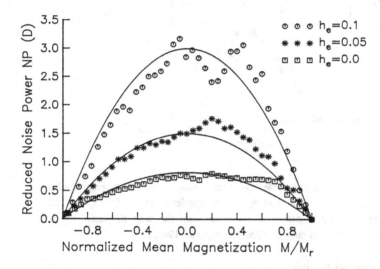

Figure 6.1. Calculated noise power as a function of $\overline{m}$ ($\overline{m}$ changes from +1 to −1) for various exchange coupling strengths. The solid curves are the fitting parabolic functions: $s(0)\times(1-\overline{m}^2)$.

Figure 6.2 shows the calculated noise power as a function of the external field for $h_e = 0$ and $h_e = 0.05$. The figure shows another important finding of this modeling study: The noise power follows the field derivative of the remanent hysteresis curve which is plotted in the figure for both the exchange coupled and nonexchange coupled cases. Further more, the normalization factor g for fitting the total noise power to the field derivative of the remanent hysteresis curve is the same for both the exchange coupled and nonexchange coupled cases: $g = 7.7$. The noise power as a function of the spatially uniform external erasure field exhibits a linear relation with the differentiation of the remanent hysteresis curve with respect to the field. Such linear relation holds for films with different intergranular exchange coupling strengths and appears to be universal for all planar isotropic films. It should be understood in the sense that the magnetization process governs both the remanent hysteresis curve and the spatial variations of the irregular magnetization patterns.

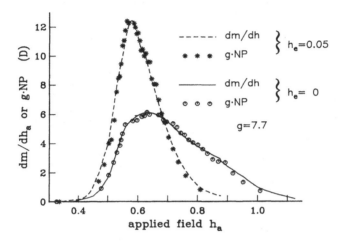

Figure 6.2. Calculated noise power as a function of the external applied field for $h_e = 0$ (stars) and $h_e = 0.05$ (circles). The solid ($h_e = 0$ and dashed ($h_e = 0.05$) curves are corresponding the field derivative of the remanent hysteresis curves. (Zhu and Bertram 1991a)

6.3. Experimental Studies

Recently, Hsu et al. have conducted an experimental study to confirm the linear relationship between the power of reverse *dc* erase noise and the field derivative of the remanent hysteresis curve (Hsu et al. 1991). The experiment was carefully designed to eliminate the demagnetizing fields in the reverse *dc* erasure process so that the noise power as a function of the external field can be obtained. *In situ* measurements of the remanent hysteresis curves were performed at an effectively demagnetization field free condition for obtaining the corresponding remanent hysteresis curve. A large scale erasing head with gap-length $g_h \approx 3mm$ and head-medium spacing $d_h \approx 0.5mm$ was utilized. For reverse *dc* erasure with the large scale head, the magnetization of the medium was initially saturated along one direction of the track. Then the field polarity of the large scale head was changed and the field is applied incrementally (stepwise with very small stepsize) from zero to large field value to saturate the magnetization in the opposite direction. The poor field gradient of the large scale head together with the incremental erasure method virtually eliminates the demagnetizing field effect in the erasure process. The *in situ* measurement of the remanent hysteresis curve was performed in the following way: A saturation squarewave recording was performed with a regular recording head at a very long wave length. Then, the large scale erase head was used to incrementally erase the recorded patterns with field applied stepwise with very small step size. At each field step, read back is performed and the rms voltage value of fundamental component is measured and reduction of the magnetization level of the recorded squarewave can be calculated according to the voltage magnitude. As the erase field increases from zero to the negative saturation field, the magnetization level reduces from $+M_r$ to $-M_r$. The *in situ* remanent hysteresis curve

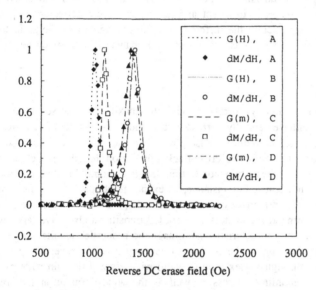

Figure 6.3 Measured noise power of reverse *dc* erasure and the field derivative of the remanent hysteresis curve as functions of external applied field
for four different thin film recording media including both low noise and
high noise media. ($G(H)$ is normalized NP_{dc}). (Hsu et al. 1991)

measured with this method showed an excellent fit with the VSM measurement, proving that the demagnetizing field effect had actually been eliminated. The study shows
that the measured noise power of the reverse *dc* erasure exhibit exactly linear relation
with the field derivative of the remanent hysteresis curve, just as the modeling study

predicted, for a variety of thin film longitudinal media, including both low noise and high noise media. Figure 6.3 shows the measured noise power and the field derivative of the remanent hysteresis curve as functions of the external applied field for four different thin film media.

Prior to this recent experimental study, two experimental investigations were conducted on the relations between the noise power and the field derivative of the remanent hysteresis curve (Silva and Bertram 1990, Tarnopolsky et al. 1991). Both experiments concluded that the noise power follows a quadratic dependence of the field derivative of the remanent hysteresis curve, suggesting that the noise arises from modulation effect due to the spatial fluctuations of either the external field or the macroscopic film hysteresis parameter such as the remanent coercivity (Bertram et al. 1986). However, in these experiments, the demagnetizing fields were very significant in both reverse *dc* erasure process and the *in situ* measurement of the remanent hysteresis curve. Experimental analysis performed by Hsu et al. has concluded that the obtained quadratic relation is incorrect and is caused by the demagnetizing fields in the measurements, specifically the difference of the demagnetizing fields in the noise measurement and the *in situ* measurement of the remanent hysteresis curve (Hsu et al. 1991).

6.4. Discussions

Transition noise directly correlates with the maximum reverse *dc* erase noise since a transition center essentially corresponds to the remanent coercive state. At the remanent coercive state, the reduced noise power is equal to the cross-track correlation length s. One can also define a cross-track correlation length at the center of a recorded transition and these two correlation lengths should be directly correlated, although they may not equal. For instance, in exchange coupled films, the cross-track correlation length at the remanent coercive state is large due to wide domains and the zig-zag domain patterns at the centers of recorded transitions also have large width as shown in Fig. 4.4a. The characteristic micromagnetic properties results from the intergranular exchange coupling in the film, such as magnetization reversal by expansion of reverse domains through vortex motion, yield wide zig-zag domain patterns at transition centers in a recording process as well as the wide domains at the remanent coercive state (Fig. 5.3b).

Assume that the cross-track correlation length s is not a function of the states along the remanent hysteresis loop (which is a relatively good approximation as the results from both the modeling and experimental studies show) and that the remanent hysteresis curve can be approximately expressed as a hyperbolic tangent function:

$$\overline{M}(H) = M_r \tanh\left[\frac{H-H_c^r}{\Delta H}\right] \tag{6.15}$$

where H_c^r is the remanent coercivity and $\Delta H = H_c^r(1-S_r^*)$ is the switching field dispersion. The differentiation of the remanent hysteresis curve with respect to the field becomes:

$$\frac{d\overline{M}(H)}{dH} = M_r \operatorname{sech}^2\left[\frac{H-H_c^r}{\Delta H}\right] = \frac{M_r}{\Delta H}\left[1-\tanh^2\left[\frac{H-H_c^r}{\Delta H}\right]\right] = \frac{M_r}{\Delta H}\left[1-\overline{m}^2(H)\right] \tag{6.16}$$

Hence, we have

$$NP_{dc}(H) = \frac{s\,\Delta H}{M_r}\times\frac{d\overline{M}}{dH}. \tag{6.17}$$

The reduced noise power becomes exactly proportional to the field derivative of the remanent hysteresis curve. There are two points to be stressed here. First, the results form computer modeling study show (Fig. 6.2) that the product of the cross-track correlation length s and switching field dispersion ΔH is approximately a constant, suggesting that the width of the domains at the remanent coercive state and the remanent coercive squareness are correlated. The relationship can be explained, at least qualitatively, by the fact that it is virtually the intergranular exchange coupling which results in wide domains as well as high coercive squarenesses, or narrow switching field dispersions. Second, in both the experimental and computer modeling studies, the noise power slightly deviate from the function $1 - \overline{m}^2$ and the remanent hysteresis curves also show some deviation from the hyperbolic tangent function, however, the reduced noise power still follows the field derivative of the remanent hysteresis loop. The linear relation between the noise power and the field derivative of the remanent hysteresis curve has been found to be true for all the longitudinal film media which we have measured. This relation could be fundamental and universal for thin longitudinal film media.

7. Conclusions

Computer simulation studies utilizing the dynamic micromagnetic model have provided fundamental understanding of the magnetization processes and resulting hysteresis properties in magnetic thin film recording media. Collective micromagnetization processes due to magnetostatic interactions and intergranular exchange coupling dominate the magnetization process, the hysteresis properties and the recording performance and the noise behavior. Simplified simulation model with incorporating the essential physical characteristic features of the magnetization process has been very useful for study phenomena occurring in large spatial scales.

230

8. References

Aoi, H., M. Saitoh, N. Nishiyama, R. Tsuchiya, and T. Tamura, "Noise characteristics in long-itudinal thin film media," *IEEE Trans. Magn.*, **MAG-22**, 895 (1986).

Arnoldussen, T.C., "Thin-Film Recording Media," *Proc. IEEE*, **74**, 1526 (1986)

Arnoldussen, T.C. and H.C. Tong, "Zigzag Transition Profiles, noise and correlations and statistics in Highly Longitudinal Thin Film Media," *IEEE Trans. Magn.*, **MAG-22**, 889 (1986).

Bak, P., C. Tang, and K. Wiesenfeld, "Self-Organized Criticality: An Explanation of $1/f$ Noise," *Phys. Rev. Lett.*, **59**, 381 (1987).

Barany, A.M. and H.N. Bertram, "Transition Noise Model for Longitudinal Thin-Film Media," *IEEE Trans. Magn.*, **MAG-23**, 1776 (1987).

Baugh, R.A., E.S. Murdock, and B.R. Natarajan, "Measurement of Noise in Magnetic Media, IEEE Trans. Magn., **MAG-19**, 1722 (1983).

Beardsley, I.A. and V.S. Speriosu, "Determination of Thin film Media Model Parameters Using DPC Imaging and Torque Measurements," *IEEE Trans. Magn.*, **26**, 2718 (1990)

Beardsley, I.A. and J.-G. Zhu, "DC-Erase Edge Noise Simulations in Thin Film Media," *J. Appl. Phys.*, **67**, 5352, (1990).

Belk, N.R., P.K. George, and G.S. Mowry, "Measurement of the Intrinsic Signal-to-noise Ratio for High Performance Rigid Recording Media," *J. Appl. Phys.*, **59**, 557 (1986)

Bertram, H.N. and J.C. Mallinson, "Theoretical Coercive Field for an Interacting Anisotropic dipole Pair of Arbitrary Bond Angle," *J. Appl. Phys.*, **33**, 1308 (1969).

Bertram, H. N., K. Hallemasek, and M. Madrid, "DC modulation noise in thin metallic media and its application for head efficiency measurements," *IEEE Trans. Magn.*, **MAG-22**, 247 (1986).

Bertram, H.N. and J.-G. Zhu, "Simulations of Torque Measurements and Noise in Thin Film Magnetic Recording Media," IEEE Transaction on Magnetics, **27**, 5043, (1991).

Brown, W.F. Jr., *Micromagnetics*, New York: Interscience, (1963).

Brown, W.F. Jr., "The Fundamental Theorem of Fine Ferromagnetic Particle Theory," *J. Appl. Phys.*, **39**, 993, (1968).

Chen, T., "The Micromagnetic Properties of High-Coercivity Metallic Thin Films and Their Effects on the Limit of Packing Density in Digital Recording," *IEEE Trans. Magn.*, **MAG-17**, 1181 (1981).

Duan, S.L., J.O. Artman, K. Hono, and D.E. Laughlin, "Improvement of the Magnetic Proper-ties of CoNiCr Thin Films by Annealing," *J. Appl. Phys.*, **67**, 4704 (1990).

Futamoto, M., F. Kugiya, M. Suzuki, H. Takano, H. Fukuoka, Y. Matsuda, N. Inaba, T. Taka-gaki, Y. Miyamura, K. Akagi, T. Nakao, H. Sawagucho and T. Munemoto, "Demonstration of 2 Gb/in^2 Magnetic Recording at a track density of 17 kTPI," *IEEE Trans. Magn.*, **27**, 5280, (1991).

Haas, C.W. and H.B. Callen, "Ferromagnetic Relaxation and Resonance Line Widths," *Magnetism*, Edited by Rado and Suhl, **I**, 449 (1963).

Hughes, G.F., "Magnetization Reversal in Cobalt-Phosphorus Films," *J. Appl. Phys.*, **54**, 5306 (1983)

Hsu, Y.M., J.-G. Zhu, J.M. Sivertsen, and J.M. Judy, "Mechanism of Reverse DC Erase Noise in Thin Film Media," submitted to J. Magn. Magn. Matl. (1991)

Johnson, K.E., P.R. Ivett, D.R. Timmons, M. Mirzamaani, S.E. Lambert and T. Yogi, "The Effect of Cr Underlayer Thickness on Magnetic and Structural Properties of CoPtCr Thin Films," *J. Appl. Phys.*, **67**, 4686 (1990).

Johnson, K.E., E. Wu, J.-G. Zhu, and D. Palmer, "Medium Noise Reduction Through Head-Disk Spacing Reduction, submitted to IEEE Trans. Magn. (1992).

Mansuripur, M. and T.W. McDaniel, "Magnetization Reversal Dynamics in Magneto-Optic Media," *J. Appl. Phys.*, **63**, 3831 (1988).

Mansuripur, M. and R. Giles, "Simulation of Magnetization Reversal Dynamics on the Connection Machine," *J. Appl. Phys.*, **67**, 5555 (1990).

Madrid, M. and R. Wood, "Transition Noise in Thin Film Media," *IEEE Trans. Magn.*, **MAG-22**, 892 (1986).

Mayo, P.I., K.O'Grady, P.E. Kelly, and J. Cambridge, I.L. Sanders, T. Yogi and R.W. Chantrell, "A Magnetic Evaluation of Interaction and Noise Characteristics of CoNiCr Thin Films," *J. Appl. Phys.*, **69**, 4733 (1991).

Miles, J.J. and B. Middleton, "The Role of Microstructure in Micromagnetic Models of Longitudinal Thin Film Magnetic Media," *IEEE Trans. Magn.*, **26**, 2137 (1990).

Moon, R.M., "Internal Structure of Cross-Tie Walls in Thin Permalloy Films through High-Resolution Bitter Techniques," *J. Appl. Phys.*, **30**, 82S (1959).

Nguyen, T.A., I.R. McFadyen, and P.S. Alexopoulos, "Microscopic Studies of the Magnetization-Reversal Process in Co-Alloy Thin Films," *J. Appl. Phys.*, **67**, 4713 (1990).

Silva, T.J. and H. N. Bertram, "Magnetization fluctuations in uniformly magnetized thin-film recording media," *IEEE Trans. Magn.*, **26**, 3129 (1990).

Stoner, E.C. and E.P. Wohlfarth, "A Mechanism of Magnetic Hysteresis in Hetrogeneous Alloys," *Phil. Trans. Roy. Soc.*, **A240**, 599 (1948).

Shtrikman, S. and D. Treves, "Micromagnetics," *Magnetism*, Edited by Rado and Suhl, III, 395 (1963)

Tanaka, H., H. Goto, N. Shiota, and M. Yanagisawa, "Noise Characteristics in Plated CoNiP Film for High Density Recording Medium," *J. Appl. Phys.*, **53**, 2576 (1982).

Tarnopolsky, G.J., H.N. Bertram and L.T. Tran, "Magnetization Fluctuations and Characteristic Lengths for Sputtered CoP/Cr Thin-Film Media," *J. Appl. Phys.*, **69**, 4730 (1991)

Victora, R.H., "Quantitative Theory for Hysteretic Phenomena in CoNi Magnetic Thin Films," *Phys. Rev. Lett.*, **58**, 1788 (1987a).

Victora, R.H., "Micromagnetic Predictions for Magnetization Reversal in CoNi Films," *J. Appl. Phys.*, **62**, 4220 (1987b)

Yogi, T., G.L. Gorman, C. Hwang, M.A. Kakalec and S.E. Lambert, "Dependence of Magnetics, Microstructures and Recording Properties on Underlayer Thickness in *CoNiCr/Cr* Media," *IEEE Trans. Magn.*, **24**, 2727 (1988).

Yogi,T., C. Tsang, T.A. Nguye, K. Ju, G.L. Gorman and G. Gastillo, "Longitudinal Media for 1 Gb/in^2 Areal Density," *IEEE Trans. Magn.*, **26**, 2271 (1990)

Zhu, J.-G., *Interactive Phenomena in Magnetic Thin Films*, Ph.D. thesis in physics, University of California at San Diego (1989).

Zhu, J.-G., "Noise of Interacting Transitions in Thin Film Recording Media," *IEEE Trans. Magn.*, **27**, 5040, (1991).

Zhu, J.-G. and H. N. Bertram, "Micromagnetic Studies of Thin Metallic Films," *J. Appl. Phys.*, **63**, 3248 (1988a).

Zhu, J.-G. and H. N. Bertram, "Recording and Transition Noise Simulations in Thin Film Media," *IEEE Trans. Magn.*, **MAG-24**, 2706 (1988b).

Zhu, J.-G. and H. N. Bertram, "Magnetization Reversal in CoCr Perpendicular Thin Films," *J. Appl. Phys.*, **66**, 1291 (1989).

Zhu, J.-G. and H. N. Bertram, "Study of Noise Sources in Thin Film Media," *IEEE Trans. Magn.*, **26**, 2140 (1990).

Zhu, J.-G. and H. N. Bertram, "Self-Organized Behavior in Thin Film Recording Media," *J. Appl. Phys.*, **69**, 4709 (1991a).

Zhu, J.-G. and H. N. Bertram, "Magnetization Reversal and Domain Structures in Thin Film Recording Media," *J. Appl. Phys.*, **69**, 6084 (1991b).

Zhu, J.-G. and H. N. Bertram, "Magnetization Structures in Thin Film Recording Media," *IEEE Trans. Magn.*, **27**, 3553 (1991c).

CHAPTER 7

RECORDING SYSTEMS CONSIDERATIONS OF NOISE AND INTERFERENCE

EDGAR M. WILLIAMS
READ-RITE Corporation
Milpitas, CA 95035

1. Introduction

Noise and interference create ambiguity in detecting the location of signal pulses in digital magnetic recording systems. Because peak detectors make binary decisions about the presence or absence of these pulses within a precisely defined timing window, motivation arises for understanding how system error rates depend on random and systematic influences. This chapter identifies sources of noise and interference in digital recording systems, discusses the impact of filters and equalizers on signal and noise, and relates these considerations to peak detection in the time domain. The amount of random time jitter induced by noise is viewed relative to systematic time shifts arising from intersymbol interference at read-back, small time shifts which occur when writing data, and other perturbations such as those from adjacent data tracks or unerased residuals in the guardbands between tracks. This chapter concludes with a quantitative discussion of system error rates and the dependence of rate upon head, preamplifier, and medium noise in the presence of various levels of interference of systematic origin.

2. Sources of Noise

In a well-designed digital recording system, the major contributions to noise arise from three sources: the recording head, preamplifier, and recording medium. Because these sources are independent and uncorrelated, the total noise is described by the relation

$$N_t = \sqrt{N_h^2 + N_e^2 + N_m^2} \tag{1}$$

where the subscripts h, e, and m refer to head, electronics (preamp), and recording medium, respectively. This relation is valid when noise sources are Gaussian, white, and additive. A "white" noise is one having constant energy per unit bandwidth (the noise spectral density is a constant); preamps normally exhibit white noise, whereas heads and media show spectral peaking at some frequency. In these cases, a non-uniform spectral density is integrated over the system bandwidth, and noise density

234

is expressed by the equivalent average over this bandwidth; this is the "white" value.

2.1 Head Noise

The noise behavior of recording heads can be approximately modeled as lumped-parameter parallel resonant RLC circuits; the real part of the network impedance is a frequency dependent noise source. Figure 1 shows a circuit diagram describing the impedance behavior of ferrite, Metal-In-Gap (MIG) ferrite, thin film inductive, and magneto-resistive (MR) recording heads connected to a preamplifier. Alternative circuits can be constructed.

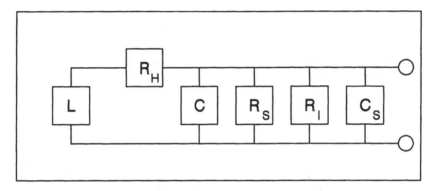

Figure 1: Equivalent Circuit for Recording Heads

R_H is the DC resistance of the coil or MR sense element, L is the total inductance of the coil and leads, C is the shunt capacitance of the element, R_S is a resistance which represents damping effects, R_I is the preamp differential input resistance, and C_S is the preamp input capacitance. Eddy currents and domain wall losses are responsible for damping in metallic thin film and ferrite heads, respectively, so R_S must be understood as a phenomenological artifact; it is not a *real* resistor. Recording heads are normally connected to preamplifiers with small diameter (44 to 48 gauge) twisted-pair wires whose length affects the network behavior as a function of frequency. The RLC electrical properties of twisted-pair wires (per centimeter of length) are approximately 0.6 ohm, 20 nanohenry, and 0.8 picofarad, respectively. For noise calculations, wire RLC values may be added directly to the head values. Experimental results for a variety of heads are given in Table 1. The effective lumped-parameter values for R_H, R_S, L, and C are determined with an impedance analyzer; f_R is the resonant frequency of the head with a wire length of about 4 cm (circuit loading by the preamp is not included here.)

Table 1: Electrical Properties of Inductive Heads

Head Type	$R_H(\Omega)$	$R_S(\Omega)$	L(nH)	C(pF)	f_R(MHz)
TFH (30-turn)	31.0	292	475	5.2	101.3
TFH (42-turn)	45.0	417	825	5.0	78.4
MIG (34-turn)	4.4	2805	1580	5.0	56.8
Mini-Composite	6.0	3410	4200	5.2	33.9
Mini-Monolithic	6.0	5410	14000	6.0	17.4

If a head is loaded by a preamplifier differential input resistance (R_I) and input capacitance (C_S), the real part of the total impedance is

$$Re[Z] = \frac{R_p}{D}[(R_H)(R_H+R_p) + (\omega L)^2],\qquad(2)$$

where the terms R_p and D are defined by the relations

$$R_p = \frac{R_S\ R_I}{R_S+R_I}\qquad(3)$$

$$D = [R_H+R_p-J(\omega^2)]^2+[F\omega]^2\qquad(4)$$

and F and J are given by

$$F = L + [R_H\ R_p(C+C_S)],\qquad(5)$$

$$J = R_p\ L\ (C+C_S).\qquad(6)$$

When a head is loaded by a preamp (assume $R_I = 750\Omega$ and $C_S = 20$ pF), Re[Z] changes substantially; Figure 2 is a plot for the heads in Table 1, and resonant frequencies are reduced by the input capacitance of the preamp.

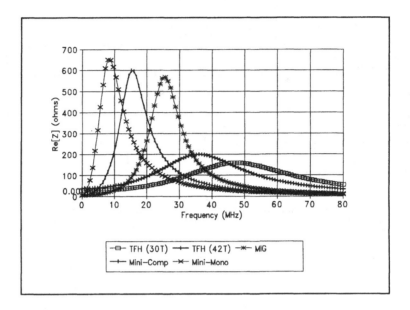

Figure 2: Re[Z] (Ω) vs Frequency (MHz)

The inductance and capacitance of MR thin film heads are normally dominated by the characteristics of the twisted-pair wires attached to the sense element. Most MR heads are fabricated with NiFe thin films 200Å to 400Å thick with a sheet resistance in the range of 12 to 6 ohms/square; an element 300Å thick, 5 μm wide, and 2.5 μm high would have a DC resistance of about 16 Ω. Twisted-pair leads 5 cm long would add about 100 nH of inductance, 4 pF of capacitace, 3 Ω of resistance, and the resonant frequency of the circuit would be over 200 MHz.

Head noise voltage is found with Nyquist's Theorem (Nyquist, 1927) for the root-mean-square e.m.f. arising from thermal agitation of electrons in the resistive component of impedances. This is given by the relation

$$e_n - \sqrt{4kT \, R \, NBW}, \tag{7}$$

where k is Boltzmann's Constant (1.3805×10^{-23} Joule/°K), T is the absolute temperature (°K), and R is the average noise impedance (Ω) over the noise bandwidth NBW (Hz) of the system. (More will be said about NBW later in this chapter.) Inspection of Figure 2 reveals the noise impedance (Re[Z]) of thin film heads is dominated by DC resistance in the frequency range below 20 MHz, whereas ferrrite and MIG head noise impedances are dominated by coil inductances and low resonant frequencies.

2.2 Preamplifier Noise

In typical low velocity applications, rigid disk recording systems operate at signals in the range of 150 to 500 microvolts (peak-peak), thus preamplifiers require low noise designs. The equivalent Gaussian (white) noise voltage of preamps arises from thermal noise across the transistor base resistance and from current shot noise flowing in the base (Horowitz and Hill, 1980, p.291). The base resistance noise decreases with current while shot noise increases, so the operating current is normally optimized to produce minimum noise for a known source impedance. The noise behavior of recording preamps is conventionally expressed as a white gaussian noise spectral density (NSD) arising from thermal and shot noise sources within the integrated circuit. Commercially available IC preamp data sheets specify voltage NSD only, which ranges from roughly 0.5 to 2.0 nV/√Hz for devices used with thin film and ferrite heads. Current noise (NSD ≈ 5 pA/√Hz) is important in ferrite and MIG head applications, however it is unusual to find a separate specification for this noise source. In future data sheets, engineers will probably find independent specifications for voltage and current noises in recording preamplifiers. Since head and preamp noise sources are uncorrelated, the total noise is computed on a root-mean-square basis, and a comparison of head and preamp noise levels will help quantify useful directions for noise reduction. For a noise bandwidth of 20 MHz, the average noise resistance of the 30-turn thin film head in Table 1 is about 35 ohms, so the noise voltage estimated from Equation 7 would be about 3.5 microvolts (RMS); this is equivalent to a NSD of 0.78 nV/√Hz. A preamp of equal NSD would also generate an equal noise voltage, so the total noise from head and preamp would be 5.0 microvolts (RMS). If there was a requirement to reduce the total NSD by a factor of two (a 6 dB improvement in signal-to-noise ratio), head and preamp noises would each have to drop by a factor of two.

2.3 Medium Noise

The noise behavior of thin film recording media is quite different from particulate media: time-averaged medium noise level normally increases with recording density in continuous metallic thin films, and decreases somewhat when recording at high density in particulate films. There is very little noise in DC-erased thin films, while noise is maximized in DC-erased particulate media. In either case, medium noise is related to highly localized (spatial) variations in the magnetization level or in the position of a transition center. These variations in magnetization produce variations in magnetic flux which are sensed as noise voltages at the output terminals of recording heads. Because medium noise is spatially related, it can be extracted from thermal noise by time-domain averaging techniques. In thin films, the noise resides in the location of a transition center, thus it is correctly called "transition jitter." In particulate media, noise is normally associated with fluctuations in the local average magnetization as summed over a volume represented by the track width, written depth, and length between transitions. The theory of medium noise and its measurement receive detailed coverage in the other chapters of this book. (Also see Nunnelley et al, 1987; Baugh et al, 1983; Arnoldussen and Tong, 1986.)

Thin film medium noise is measured with transitions written at a specified density; when viewed with a spectrum analyzer, the noise floor increases with increasing density, and the average noise density over the system bandwidth is normally determined at each recorded density. The data in Table 2 are obtained at 3.33, 5.00, and 6.67 MHz, corresponding to linear densities of 14, 21, and 28 Kfci (kilo flux changes per inch) for each of 8 thin film rigid disks. A 24-turn thin film head with track width of 17 μm, 3.0 μm thick poles, 0.5 μm gap, and flying height of 0.25 μm at 12.0 meters/sec velocity was used to obtain data from each disk. The RMS head and preamp NSD is 1.1 nV/$\sqrt{}$Hz; the system filter is a 5-pole Bessel with -3 dB noise bandwidth of 15 MHz. In all cases, the recorded signal is nearly sinusoidal, so peak-to-peak signal is $2\sqrt{2}$x(RMS signal). Medium noise scales with the readback signal level, thus a signal/noise ratio dominated by medium noise must be improved by addressing the manufacturing process or the materials used in the magnetic medium.

Table 2: Thin Film Media Signal and Noise Measurements

Disk Type and H_c	Density (Kfci)	Signal (μV, pk-pk)	Total NSD (nV/$\sqrt{}$Hz)	SNR (pk-pk/RMS)
Plated	14.0	308	2.00	39.9
NiCoP	21.0	238	2.66	23.2
800 Oe	28.0	105	2.92	9.3
Sputtered	14.0	368	2.74	34.8
CoPt	21.0	356	3.59	25.7
950 Oe	28.0	204	4.39	11.9
Sputtered	14.0	379	2.00	48.9
CoCrTa	21.0	368	2.83	33.7
950 Oe	28.0	167	3.59	11.9
Plated	14.0	308	1.78	44.7
CoNiP	21.0	322	2.35	35.4
950 Oe	28.0	164	2.85	15.0
Sputtered	14.0	337	1.88	46.1
CoCrTa	21.0	371	2.47	38.8
1050 Oe	28.0	201	3.99	13.0
Plated	14.0	277	1.66	43.0
CoNiP	21.0	280	1.91	37.9
1050 Oe	28.0	175	2.62	17.3
Plated	14.0	308	1.70	47.0
CoNiP	21.0	351	2.07	43.8
1200 Oe	28.0	218	2.52	22.3
Plated	14.0	308	1.35	58.8
CoNiX	21.0	365	1.39	67.9
1200 Oe	28.0	223	1.61	35.9

The entries in Table 2 reveal some of the experimental complexities of noise analysis: signal level depends on disk properties, flying height, and linear density, while total noise is the RMS sum of head, preamp, and disk transition jitter. The signal at high density (28 Kfci) improves with increasing coercivity (H_c), while noise is sensitive to the magnetic alloy and to subtle, but undisclosed processing considerations. Medium transition jitter can be extracted from the total noise by using Equation (1). These data are from disks which were commercially available in 1986. By 1989, most rigid media were in the 1200 to 1300 Oe range, head flying heights dropped to about 0.15 μm, and noise data at 8.43 m/sec velocity showed important improvements in transition jitter: total NSD was about 1.5 to 1.9 nV/$\sqrt{}$Hz at 24 Kfci with head and preamp NSD held at 1.1 nV/√Hz. Signal levels at 24 Kfci (4 MHz) were in the range of 230 to 300 μV (pk-pk). At the time of this writing (1991), disk coercivity has increased to 1450 Oe, flying height has dropped to 0.10 μm, density has increased to about 40 Kfci, velocity has decreased to about 6.5 m/sec, and transition jitter is on par with head-preamp NSD of about 0.8 to 1.0 nV/√Hz.

3. Filtering, Equalization, and Noise Bandwidth

Because digital recording systems function in the time domain, noise creates ambiguity in detecting the location of signal pulses, so the system error rate can rise to unacceptable levels unless filtering circuits are employed to reduce the influence of noise. Low pass (LP) filters are normally used to reduce noise in recording systems, however an interesting trade-off exists when optimizing filter designs for minimum error rate. When filter bandwidth is reduced, noise reduces, but the LP filter impulse response broadens the filtered signal pulsewidth (with Butterworth LP filters, pulsewidth may narrow under some conditions of medium velocity and filter bandwidth). At high densities, broad pulses overlap more than narrow pulses, so intersymbol interference (ISI) increases and signal amplitude drops. When signal loss overtakes noise reduction, further narrowing of filter bandwidth degrades the error rate, thus an optimum bandwidth exists. Historically, filters have been designed and analyzed in the frequency domain, but with pulse detection systems a time domain approach gives direct and immediate insight, and is therefore preferred over the frequency domain approach. For this reason, the impulse response function of filters and equalizers is emphasized in this chapter. Indeed, in linear network theory the impulse reponse is a complete description of any linear circuit, but in the frequency domain, amplitude and phase reponses must be treated separately (Blinchikoff and Zverev, 1987.)

The distinction between filters and equalizers is somewhat false because equalizers are a class of filters. In digital recording system terminology, equalizers are conventionally understood as filters whose impulse responses are designed to change signal pulse shapes in certain desirable ways. With peak detection channels, equalizers are used to reduce the pulsewidth of readback pulses; electronic slimming reduces the amount of pulse overlap, reduces ISI, increases signal at high densities, and increases the total noise. Figure 3 is a plot of the impulse reponses of a Bessel LPF (5 poles, -3 dB bandwidth = 10 MHz) and an equalizer with 5 Bessel poles (-3 dB at 5 MHz) and 2 zeroes giving 8 dB of boost at 5 MHz. Figure 4 shows the computed impact of these circuits on the readback pulse from a thin film head flying at 0.125 μm above a 1400 Oe thin film magnetic medium. Pole and gap geometry are P1/G/P2 = 3.5/0.35/3.5 microns, and the medium velocity is 7.0 m/sec. The first pulse is unfiltered, the second is filtered by the 10 MHz Bessel LPF, and the third is slimmed by the equalizing network (Kost and Brubaker, 1981).

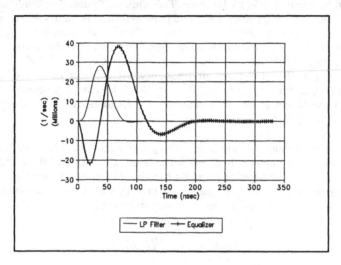

Figure 3: Impulse Response Functions

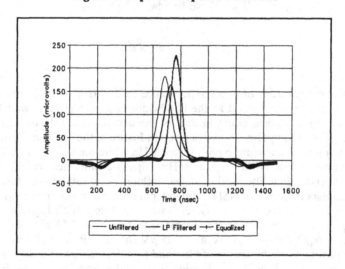

Figure 4: Unfiltered, Filtered, and Equalized Pulses

The unfiltered pulse amplitude is 182 μV and the pulsewidth PW50 is 109 nsec. Filtering with the linear phase (5 pole Bessel) LP filter drops the amplitude to 164 μV and broadens PW50 to 112 nsec. The equalized pulse amplitude and PW50 are 225 μV and 91.4 nsec, respectively, and pulse response is clearly sensitive to exact details of the equalizer impulse response. In "frequency domain" language, the equalizer is specified by the number and types of poles, the -3 dB voltage cut-off frequency (bandwidth) before adding high frequency boost, and the amount of voltage boost (in dB) at the cut-off frequency. At a linear density of 36.3 Kfci (equivalent to a signal frequency of 5.0 MHz) the pulses overlap significantly; the 5 MHz signal amplitude is 297, 261, and 484 μV (peak-peak) respectively, for the unfiltered, filtered, and equalized pulses.

Boosting the signal also increases noise, and Parseval's Theorem is useful in calculating the noise bandwidth of any filter or equalizer (Blinchikoff and Zverev, 1987, p.32). If h(t) is the impulse response function of a two-port network and H(ω) is the frequency domain Fourier transform of h(t), where ω = 2πf is the radian frequency, then the total signal energy is proportional to

$$\int_{-\infty}^{+\infty} |h(t)|^2 dt = \frac{1}{2\pi} \int_{-\infty}^{+\infty} |H(\omega)|^2 d\omega . \tag{8}$$

This is Parseval's Theorem. The "Noise Bandwidth" (NBW) is

$$NBW = \frac{1}{2} \int_{0}^{+\infty} |h(t)|^2 dt, \tag{9}$$

and simplifies comparisons of noise performance for filters and equalizers. In other words, NBW can be understood as the bandwidth of a perfect brick-wall LP filter which gives a noise voltage equal to that of the actual filter. (Horowitz and Hill, 1980, p.306.) For the impulse responses given in Figure 3, the 10 MHz 5 pole Bessel LP filter has a noise bandwidth NBW = 10.4 MHz, while the equalizer (5 MHz cut-off, 8 dB boost) yields NBW = 30.6 MHz. Specific numerical calculations are helpful in understanding filter and equalizer influences on total noise. Assume that the white Gaussian equivalent NSD's of head, preamp, and medium are 0.8, 0.8, and 1.0 nV/$\sqrt{}$Hz, respectively when measured with a LP filter. With Equation (1), the three independent noise sources are combined to give N_t = 1.51 nV/$\sqrt{}$Hz, yielding noise voltages $N_t\sqrt{NBW}$ = 4.87 μV and 8.35 μV for the LP filtered and equalized cases, respectively. With this choice of filter and equalizer designs, the signal-to-noise ratio *improves* by 0.68 dB with equalization because the equalized cut-off frequency (5 MHz) is set at the signal frequency, and boost is not excessive (8 dB).

4. Noise in the Time Domain

Signal, noise, and error probability for peak detection are closely related. Peak detection involves time domain differentiation of the signal-plus-noise and finding zero-crossing locations whose uncertainty depends upon the amount of differentiated noise. When noise amplitude probability density is Gaussian, the soft error rate for Lorentzian pulses is approximately described by (see equation (14))

$$E.R. - 1 - erf(z) - erfc(z) ,$$
$$where \ z - SNR\frac{T_W}{PW50} . \tag{10}$$

Erf(z) is the error function, erfc(z) is the complementary error function, SNR is the signal (peak-to-peak) to noise (RMS) ratio after differentiation, PW50 is the time-domain undifferentiated pulsewidth, and T_W is the center-to-edge time of the detector window. Influences of filters, equalizers, and differentiators are included. (Hughes and Schmidt, 1976; Katz and Campbell, 1979; Williams, 1990; Hoagland and Monson, 1991.) In frequency domain analyses the signal is sinusoidal, so the soft error rate (error probability) is written

$$E.R. - erfc(2\pi f \ SNR \ T_W) . \tag{11}$$

In this form, SNR is (RMS/RMS) and PW50 does not appear because the signal slope is $(d/dt)(V_0 sin(2\pi ft) = 2\pi f V_0 cos(2\pi ft)$. Since digital recording systems operate in the time domain, Equation (10) is more general for estimating time domain jitter. Selected values of z and erfc(z) are given in Table 3 (Abramowitz and Stegun, 1965, p.299, ¶ 7.1.28.)

Table 3: Error Rate

z	$\log_{10} erfc(z)$
2.75	-4.0
3.13	-5.0
3.46	-6.0
3.78	-7.0
4.07	-8.0
4.35	-9.0
4.57	-10.0

Many recording systems are specified to operate with a soft error rate less than $1x10^{-10}$, so Equation (10) can be solved for the timing window (center-to-edge) which would just meet that criterion. In the absence of systematic time shifts arising from ISI or writing phenomena, the timing window would be consumed by noise-induced jitter, and this time may be written (Williams, 1990)

$$T_W - T_{WSNR} = \frac{4.57 \ PW50}{SNR} \ . \tag{12}$$

Figure 5 compares calculations and experiments with two thin film disks; SNR is in units of (p-p)/RMS and PW50 is 140 nsec in both cases. The thin film head, preamp, and LP filter remained unchanged for the experiments, and the difference in SNR arose from differences in disk transition jitter.

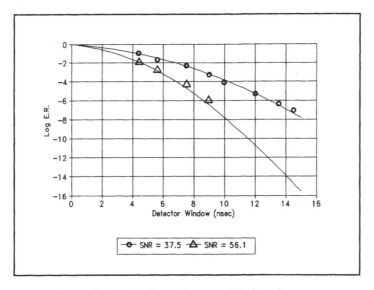

Figure 5: Log_{10} E.R. vs Detector Window Size

The experimental points agree well with the complementary error function curves, so head, preamp, and transition noise may be treated as additive white Gaussian sources. Some workers discuss error rate in terms of the normal probability distribution P(x) (with mean m and standard deviation σ) which is closely related to the error function erf(z) and erfc(z) = 1 - erf(z). P(x) is given in many sources (Abramowitz and Stegun, 1965), and its relation to erf(z) is

$$P(x) = \frac{1}{2}\left[1 + erf\left(\frac{x-m}{\sigma\sqrt{2}}\right)\right] . \tag{13}$$

With m = 0, $P(6.36) \approx erfc(4.57) \approx 1\times10^{-10}$, so the arguments must be scaled by x = z/2.

5. Importance of Noise Sources Relative to Systematic Time Shifts

5.1 Pattern-Induced Peak Shift (ISI)

While the main concern of this book is "noise," it is necessary to appreciate noise in the context of other influences which impact peak detection and system error rate. Since peak detection involves time domain differentiation of the signal, the peak location of any given pulse will be shifted early or late by the influence of adjacent pulses; this is called intersymbol interference (ISI) or pattern-induced shift. The amount of time shift depends on the *slope* of the interference, and this type of peak shifting is systematic in origin, because it depends on the shape of isolated pulses and the written data pattern. First-order estimates of ISI peak shift can be obtained by assuming the pulse shape is described by an uncomplicated mathematical equation. The Lorentzian pulse is a useful example, and it is given by the relation

$$e(t) = \frac{E_o}{1 + \left(\dfrac{2t}{PW50}\right)^2} \tag{14}$$

where E_o is the base-peak pulse amplitude, PW50 is the pulsewidth at the 50% level, and t is time. For small time displacements near the peak, this equation further simplifies to a parabolic form,

$$e(t) \approx E_o\left[1 - \left(\frac{2t}{PW50}\right)^2\right] . \tag{15}$$

246

If an interference voltage of the form $V_i = Kt + b$ is added to the signal of interest, the peak position is shifted; the new peak location is found by solving the equation

$$\frac{d}{dt}\left[e(t) + V_i\right] = 0 \tag{16}$$

and the resultant shift (T_P) is

$$T_P = \frac{K\ PW50^2}{8E_o}. \tag{17}$$

Pattern shifting is thus directly proportional to the slope K (volts/sec) of the interference, and scales with $PW50^2$. The pulses shown in Figure 4 (from thin film heads) are more complicated than simple Lorentzians because of the presence of undershoots arising from the thickness of finite poles, and filter/equalizers alter the pulse shape as well. Figure 6 shows early and late peak shifting of about $T_P=-7.5$ nsec and $T_P = +7.3$ nsec, respectively from ISI at 36.3 Kfci for two pulses (a "dipulse") of a thin film head. The difference in early and late shifting arises from pulse shape asymmetry induced by the LP filter. In this computed result, the pulses are nearly Lorentzian in shape down to a level of about 25% of the maximum pulse amplitude.

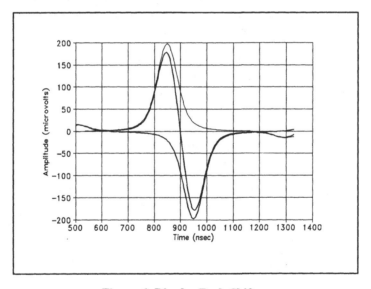

Figure 6: Dipulse Peak Shift

247

Dipulses are the simplest pattern for a discussion of ISI, however recording with pseudo-random data patterns reveals a *distribution* of peak locations which depends on details of the recording code, linear density, and filtered (or equalized) isolated pulse shape. Figure 7 shows two computed peak shift histograms for random patterns encoded with a run-length-limited RLL (1,7) code; the first probability density is for a thin film head, and the second histogram is the same pulse with undershoots eliminated. Notice how pulse undershoots can spread the distribution of T_P. The density is 36.3 Kfci (5 MHz at 7.0 m/sec) and the isolated pulse amplitude is E_o = 154 μV with PW50 = 119 nsec.

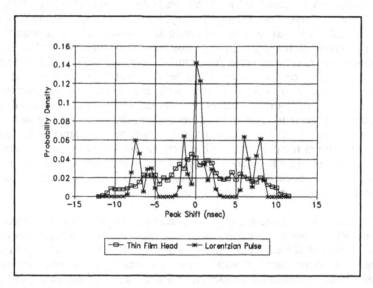

Figure 7: Peak Shift (T_P) for Random RLL(1,7) Data Patterns

The noise-induced jitter T_{WSNR} is estimated using Equation 12: if the filtered signal (at 36.3 Kfci) is 234 μV (p-p), and total noise (Equation 1) is 4.47 μV (RMS), then T_{WSNR} = 10.4 nsec at the 1×10^{-10} level. At 5 MHz signal frequency, the center-to-edge width of the detector window is 25.0 nsec for a RLL (1,7) code, so noise-induced jitter consumes 42% of the half-window, while T_P is distributed over the interval from -11 to +11 nsec, or about 44% of the half-window. In other words, total noise consumes roughly the same window as systematic pattern shift.

5.2 Write-Induced Peak Shift (Overwrite)

Write-induced shift is an additional source of peak shift which must be considered in magnetic recording. When transitions are written, the medium just entering and leaving the gap region influences the position of the flux reversal, and the direction of this systematic influence depends on the polarity of the incoming magnetization relative to the outgoing magnetization. If the gap field reverses the direction of the incoming magnetization, the transition is shifted late (downstream), while a non-reversing transition is shifted early (upstream) (Fayling et al, 1984; Rosscamp and Curland, 1988; Tsang and Tang, 1988; Williams, 1990.) In addition to this spatial shift, there exists a time-domain shift whose source is found in the switching behavior of the write-driver transistor circuit attached to the recording head. Manufacturers of write-driver/preamp I.C. chips normally specify the amount of "time asymmetry" or "pulse-pairing" to be expected from their products; it is usually in the range of 0.5 to 2 nsec with the driver output shorted. The head/medium and write-driver sources of write-induced shift are independent and uncorrelated, thus early and late arrival times add or subtract depending on the exact phase relation between the write-driver flip-flop and the remanence polarity of the recording medium as it moves into the head gap region. In either case, the early and late shifting is equivalent to *phase modulation* of the readback signal, and because this modulation has been historically measured in the frequency-domain, its time-domain nature is often misunderstood or lost.

Frequency domain analysis of a phase-modulated signal leads to a spectral line at the modulation frequency and to rich spectra of sidebands and harmonics of the signal of interest. When evaluating the effectiveness of saturated writing in digital recording, it is conventional to write a low frequency (LF) signal, measure the amplitude of the fundamental frequency, and write a high frequency (HF) signal directly over the LF pattern *without DC erasure*. The HF signal is then analyzed in the frequency-domain, and the amplitude of the LF spectral line is normalized to the initial LF fundamental level and expressed as a *residual* level (in dB) of the original LF signal. This is called "Overwrite (OW)" (Hoagland and Monson, 1991, p.106).

$$OW = 20 \log_{10} \left(\frac{LF\ Residual}{LF\ Original} \right) . \qquad (18)$$

When transitions are shifted by small amounts (which is normally the case in write-induced peak shift) a high frequency signal suffers phase shifting by an amount

$$\Delta\phi = 2\pi f \frac{T_{asm}}{4} = 2\pi f \, T_{WO} \qquad (19)$$

where f is the HF signal (Hz), and T_{asm} is the "asymmetry" or *difference* in time between successive pairs of pulses. The factor of 4 arises when pulses are self-clocked in a phase-locked-loop, because the detector window centers itself around the average arrival time of pulses (Hoagland and Monson, 1991, p.105), so T_{WO} is the center-to-edge time shift measured with a window margin tester. In general, there are four discrete arrival times associated with asymmetry: (1) early and (2) late transitions arising from positive and negative polarities of the recording medium, and (3) early and (4) late current transitions in the recording head arising from the write-driver circuit. This idea can be expressed (Williams, 1990) as

$$T_{asm} = \pm \, T_o \, \frac{H_c}{H_m} \pm T_{awd} \qquad (20)$$

where T_o is the 0-50% risetime of the writing field, H_c is the medium coercivity, H_m is the maximum writing field at the medium, and T_{awd} is the asymmetry of the write-driver circuit. Experiments which do not control the polarity and phase relationships between (1), (2), (3), and (4) above can suffer from problems in repeatability of measurements. When the magnitudes of T_{awd} and $T_O(H_c/H_m)$ are nearly equal, these terms may annihilate or double the total asymmetry from one test to the next, so overwrite could fluctuate significantly.

For small phase shifts, the amplitude of the LF spectral line is directly proportional to the phase shift (Robins, 1982, p.9), so the relation between OW (in dB) and time is

$$OW \, (dB) = 20 \, \log_{10}\!\left(\frac{\pi}{2}f \, R \, T_{asm}\right) \qquad (21)$$

where R is the HF/LF signal resolution. Note that OW scales directly with the logarithm of frequency, so doubling f degrades OW by 6 dB. In Figure 8, experiment (at f = 5.0 MHz and R = 0.95) is compared with Equation (21), and agreement is quite reasonable.

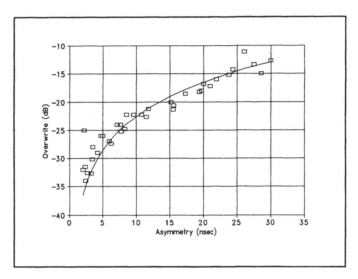

Figure 8: Overwrite and Asymmetry

Overwrite is commonly specified (in dB) while asymmetry is usually ignored, and this practice reveals how deeply frequency-domain ideas are embedded in the digital recording industry. Because peak detection works in the time-domain, there is scant justification for specifying overwrite in dB when T_{asm} and T_{WO} are readily measured. The data in Figure 8 show that T_{WO} can be held below 2.0 nsec ($T_{asm} \leq$ 8.0 nsec) if overwrite is less negative than -25 dB at f = 5 MHz. So, in the context of noise and pattern-induced peak shifts, write-induced peak shift can be held to less than 10% of the detector window, provided the sources of asymmetry are understood and addressed.

6. Adjacent Track and Guardband Interferences

Recording heads are sensitive to magnetic flux from regions within about ten microns up and down a written track, and to about three microns on either side of the head width. When data tracks are squeezed together at high densities, side-reading may introduce significant interference with the signal of interest, so this type of influence must be examined and quantified in relation to noise, pattern, and write-induced shifts of the signal pulse. Generally speaking, recording head position will vary somewhat from the intended position when writing new data or when reading previously existing tracks, so the signal will be mixed with guardband noise and interference from an adjacent track. Side-reading behavior can be characterized by profiles of signal plotted as a function of offtrack distance, and Figure 9 shows two computed offtrack profiles normalized to unity amplitude for ease of comparison. The geometrical width of the head is 7.5 μm, the written width is 8.6 μm, and side-reading broadens the electrical width by an additional amount of approximately $2S/\pi$, where S is the distance between transitions.

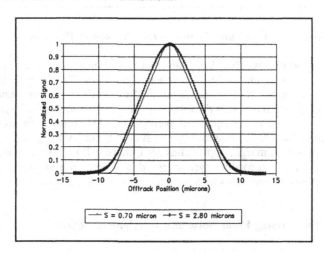

Figure 9: Normalized Offtrack Profiles

The level of adjacent track interference must be evaluated in terms of its impact on error rate and erosion of detector window margin. The experimental results in Figure 10 demonstrate this rather well: a data track 13.0 μm wide is written at Z = 0, one adjacent track is placed to the left at Z = -16.5 μm, and the medium is DC erased on the right. The consumed window (TW10) at a soft E.R. of 1×10^{-10} includes all noises and interferences, and TW10 is measured as a function of offtrack position. Figure 10 shows the signal + interferences and the consumed window (TW10).

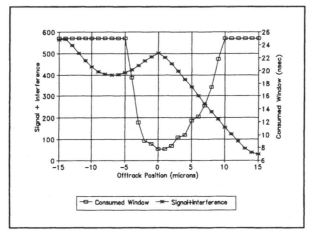

Figure 10: TW10 and Signal + Interference vs Position

As the head moves offtrack, signal reduces and interference increases, so the intrinsic error rate deteriorates until window margin is lost. The offtrack distance for zero margin is called the "Off Track Capability," or OTC. In the experiment shown in Figure 10, OTC is 5.0 μm with the adjacent track at 16.5 μm and 10.0 μm without this interference. If tracks are squeezed even closer, OTC reduces further, so track density is limited by side-reading and by the system accuracy for positioning heads. Clearly, if the on-track margin approaches zero, there will be no tolerance for head position error, so the recording system could not meet the required error rate criterion under stress conditions of temperature, vibration, or creep.

7. System Errors Arising From Noise and Interference Sources

This chapter has introduced and discussed three sources of noise (head, preamp, and medium) and three sources of interference (pattern peak shift, write-induced peak shift, and adjacent track side-reading.) These factors each contribute to problems in data integrity and unacceptable error rates. With Gaussian noise sources, systematic time shifts, and interferences, the system error rate is described by a generalization of Equation (10), namely

$$E.R. - erfc\left[\frac{SNR}{PW50}(T_W-T_P-T_{WO})\right] \qquad (22)$$

where T_W is the center-to-edge detector window, T_P is the pattern-induced shift, and T_{WO} is the write-induced shift. (The times T_W, T_P, and T_{WO} are in the positive (late) sense here. For negative (early) times, the algebraic sign is reversed.) In Equation (22), SNR includes the influence of adjacent track and guardband interferences which, within experimental accuracy, can be added on a root-mean-square basis with noise. That is, the "effective" signal-to-noise ratio is

$$SNR - \frac{Signal}{\left[(e_n)^2+(e_i)^2\right]^{0.5}} \qquad (23)$$

where e_n is the RMS noise, and e_i = (peak-peak interference/2/2). Figure 11 gives a comparison of experiment with computations based on Equations (22) and (23), and the excellent agreement shows interference is "noise-like."

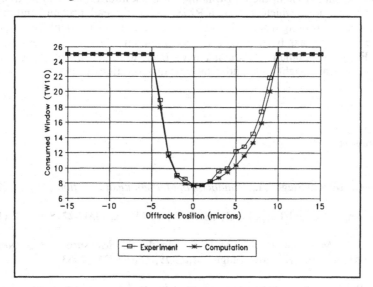

Figure 11: Experimental and Computed Profiles of TW10

In summary, digital recording systems are designed to store and retrieve data at an error rate which is acceptable to the user. With well-defined data patterns and good experimental controls, pattern-induced shifts, write-induced time shifts, and side-reading behave as systematic interferences which move the signal peak. Head, preamp, and medium noises are uncorrelated random sources which can be added together on a root-mean-square basis, although medium noise arises from spatial jitter of the zero-crossings of written transitions, whereas head and preamp noises are in the time-domain. The central thrust of this chapter has been to place all noise and interference sources on the common basis of time measurement such that comparisons of impacts on system timing and error problems are clearly portrayed.

When data patterns are encoded and randomized, or when time asymmetries of the head/medium and write-driver are randomly phased, these influences on detector window margin are also spread out and appear to be less systematic. The timing shifts from noise and interference are distributed in the early and late directions, and phase-locked-loop clocking attempts to center the detector window on the algebraic average value of total shift from all sources. In the typical forensic cases discussed in this chapter, noise sources consume approximately the same window portion as do sources of on-track interference, while write-induced shifting is typically about 5-10% of the half-window. Offtrack interference can be treated as an uncorrelated noise which adds on a RMS basis with head, preamp, and medium noises, thus window margin and offtrack capability are interdependent. Finally, digital magnetic recording has been treated in the time-domain where the physics rightly puts it. Frequency-domain treatments have been used historically, much to the detriment of clear insight and understanding of saturation recording.

8. References

Abramowitz, M. and Stegun, I., *Handbook of Mathematical Functions*, Dover, (1965).

Arnoldussen, T. C., and Tong, H. C., IEEE Trans. Magn., **MAG-22**, 889-891, (1986).

Baugh, R. A., Murdoch, E. S., and Natarajan, B. S., "Measurement of Noise in Magnetic Media", IEEE Trans. Magn., **MAG-19**, 1722-1724, (1983).

Blinchikoff, H. J. and Zverev, A. I., *Filtering in the Time and Frequency Domains*, Krieger (1987).

Fayling, R. E., Szczech, T. J., and Wollack, E. F., "A Model for Overwrite Modulation in Longitudinal Recording", IEEE Trans. Magn., **MAG-20**, 718-720, (1984).

Hoagland, A. S. and Monson, J. E., *Digital Magnetic Recording*, Wiley-Interscience (1991).

Horowitz, P. and Hill, W., *The Art of Electronics*, Cambridge Univ. Press, 1980.

Hughes, G. F. and Schmidt, R. K., "On Noise in Digital Recording", IEEE Trans. Magn., **MAG-12**, 752-754, (1976).

Katz, E. and Campbell, T., "Effect of Bitshift Distribution on Error Rate in Magnetic Recording", IEEE Trans. Magn., **MAG-15**, 1050-1053, (1979).

Kost, R. and Brubaker, P., "Arbitrary Equalization with Simple LC Structures", IEEE Trans. Magn., **MAG-17**, 3346-3348, (1981).

Nunnelley, L. L., Heim, D. E., and Arnoldussen, T. C., "Flux Noise in Particulate Media: Measurement and Interpretation", IEEE Trans. Magn.,**MAG-23**, 1767-1775, (1987).

Nyquist, H., "Thermal Agitation of Electronic Charge in Conductors", Phys. Rev., **29**, 614, (1927).

Robins, W.P., *Phase Noise in Signal Sources*, IEE Peregrinus, (1982).

Rosscamp, T. A. and Curland, N., "A Self-Consistent Model for Overwrite Modulation in Thin Film Recording Media", IEEE Trans. Magn., **MAG-24**, 3090-3092, (1988).

Tsang, C. and Tang, Y., "An Experimental Study of Hard-Transition Peakshifts through the Overwrite Spectra", IEEE Trans. Magn., **MAG-24**, 3087-3089, (1988).

Williams, E. M., "Monte Carlo Simulation of Thin Film Head Read-Write Performance", IEEE Trans. Magn., **MAG-26**, 3022-3026, (1990).

CHAPTER 8

PRACTICAL NOISE MEASUREMENTS

LEWIS L. NUNNELLEY
IBM Corporation
San Jose, CA 95193

1. Introduction

In this chapter some of the practical aspects of noise measurements will be discussed. This chapter is not intended to be an encyclopedic survey of all noise measuring techniques. Instead, selected aspects of the more common techniques are described. Hopefully the discussion will help experimentalists decide which technique is appropriate for a given goal. In addition some of the factors which influence the accuracy and precision of a given measurement are presented.

An overview of noise behavior will be given with an emphasis on those aspects of noise behavior which affect the choice of measurement technique. Generally there are two motivations for measuring noise from recording media. One is to gain more detailed knowledge of physical noise mechanisms in recording media or to measure noise in such a way to allow the accurate prediction of error rates. Usually spectrum analyzers and time recording instruments (sampling oscilloscopes and time interval analyzers) have been most useful. The second motivation is to certify media in a manufacturing environment. Usually the goal here is to develop very rapid techniques which are able to indicate when the manufacturing process has shifted. Techniques derived from both motivations are discussed in subsequent sections.

2. Physical Behavior of Noise

Every electrical resistor with finite resistance presents noise at the terminals. This effect was first reported in 1928 by J. B. Johnson (Johnson, 1928). The equation which quantifies the rms noise is:

$$V_{rms} = \sqrt{4kTRB} \tag{1}$$

where:

        k = Boltzmann's constant
        T = temperature in degrees K
        R = resistance in ohms
        B = noise equivalent bandwidth

258

The bandwidth, B, is the bandwidth through which the measurement is made and is generally wider than the 3 dB bandwidth normally used to characterize actual filters. Johnson noise has a gaussian amplitude distribution. This is illustrated in Figure 1(a). In the frequency domain Johnson noise is white (Figure 1(b)). These properties allow for a convenient measurement and description of Johnson noise.

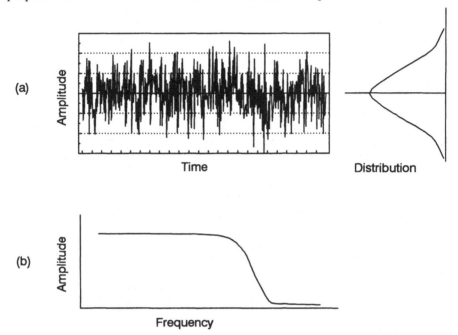

Figure 1. Random (Johnson) Noise: (a) Time domain, (b) Frequency domain.

Noise from thin film media differs in important ways from that of particulate media. One difference has to do with stationarity. Suppose a function, N(t), describes a noise voltage with respect to time. If either the mean or any higher order moment of N(t) is time (or space) dependent, then the noise is non-stationary (Bendat and Piersol, 1980). For media noise the second moment, the variance, given by:

$$\psi^2(t) = \lim_{T \to \infty} \frac{1}{T} \int_0^T N^2(t)\, dt \qquad (2)$$

is convenient to use as a test for stationarity. Johnson noise is stationary. Noise from particulate media can be assumed to be stationary; however, this assumption deserves some qualification. Noise from a particulate medium originates mostly from the particles (there can be additional noise from dispersion and modulation effects). Since there are many particles distributed randomly along a given track then to a good approximation the noise is a stationary random process. In another sense noise from particulate media is deterministic. For a head flying over the same section of the same track on a rigid disk the details of the noise would be exactly the same. In this sense there is no difference between the media signal and media noise except there is no recorded information in the noise. Therefore in a strict sense the estimate of the autocorrelation is valid as long as the track is not repeated. Practical time domain measurements of $N(t)$ usually correspond to distances much shorter than a physical track so the assumption of stationarity for particulate noise is justified.

Noise from thin film media has a substantially different behavior than from particulate media (Baugh, et al.,1983). Thin film noise depends very strongly on the state of magnetization of the film. Usually there is very little noise from a dc erased track compared to noise from a track with written transitions. The experimental procedures must therefore measure noise in the presence of signal. The noise in the presence of signal is slightly modified for particulate media but much less dramatically than for thin film media. Since thin film noise is associated with recorded transitions the variance function is not insensitive to the value of time. Therefore thin film noise is not stationary. This feature also complicates the choice of measurement schemes. The classification of noise as stationary or non-stationary is very useful when considering measurement schemes.

3. Noise Measurement in the Frequency Domain

Most frequency domain measurements are made with a spectrum analyzer (Engelson, 1989; Gallant, 1991). In this section the basic architecture of a spectrum analyzer is discussed. In addition, the more important factors which influence the performance are presented. For completeness it should be mentioned that there is another class of frequency domain measurements based on taking the Fourier transform of sampled data. These sample techniques are generally more limited in bandwidth than spectrum analyzers. Because of this limitation spectrum analyzers are generally more useful for recording media noise measurements. The discussion here is restricted entirely to spectrum analyzers.

Figure 2. shows a diagram of the basic architecture of a spectrum analyzer. The incoming signal is low pass filtered and mixed with the signal from a local oscillator. One of the resulting sidebands (usually the lower sideband) is then passed through an intermediate filter (IF). The IF functions as a narrow bandpass filter (BPF) and is labelled as such in Figure 2. The output from the intermediate filter is then presented to a log amplifier and an envelope detection circuit and this result is then presented to an output display. This model of a spectrum analyzer is greatly simplified. Actual analyzers have elaborate circuits for filtering, frequency control, and post-collection data manipulation. However, consideration of this basic model leads to an adequate understanding of the operation of a spectrum analyzer. The overall architecture is very similar to that for radio receivers and is sometimes referred to as a superheterodyne.

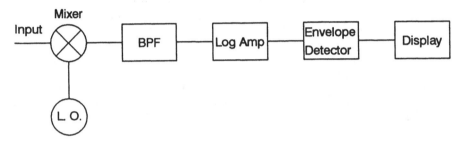

Figure 2. Spectrum analyzer Architecture, L. O. - Local Oscillator, BPF - Band Pass Filter

The performance of the intermediate filter is critical to the operation of the spectrum analyzer. These filters have to satisfy two functions. One, the width and shape of the filter (the 60 dB to 3 dB frequency ratios) must allow the resolution of closely spaced spectral components. Two, the impulse response must be relatively free of ringing and overshoot in order to optimize sweep times. The best compromise is a gaussian filter and most spectrum analyzers offer several gaussian IF filters with different bandwidths.

The envelope detection circuit following the IF responds to the peak voltage. For this reason this circuit is sometimes called a peak detector. However the display is usually calibrated so that the rms voltage is displayed. Occasionally this causes some confusion. The peak voltage of an incoming sine wave will be measured by the peak detection circuit and then the result will be multiplied by 0.707 to display the equivalent rms voltage of the incoming signal. This works fine if the incoming signal is a sine wave (or a harmonically rich signal composed of many discrete sine waves). However this automatic scaling by 0.707 gives an incorrect result when measuring broadband noise. This effect will be discussed quantitatively below. It is important to realize that the spectrum analyzer is a peak sensing voltmeter - it does not directly measure power.

Modern spectrum analyzers usually have embedded microprocessors which automatically select appropriate values of the sweep time once the user has selected values for the IF (often call the resolution bandwidth, RBW) and the frequency sweep range (often called the span). Generally the sweep time (ST) is related to the span and RBW by:

$$Sweep\ Time \propto Span/(RBW)^2 \qquad\qquad (3)$$

The selection of the RBW has a large effect on the sweep time. Generally the RBW values in the range of 10 kHz to 100 kHz give good frequency resolution of noise effects such as modulation sidebands without excessive sweep times.

There are two post detection filtering techniques which are employed in spectrum analyzers. Video filtering is additional filtering of the peak detected signal. The video bandwidth is less than the RBW and causes an increase in the sweep time. Video averaging is also available. In this case several independent sweeps are averaged. The final statistical result is similar for both filtering techniques, and the total acquisition time is approximately equivalent. However the sweep times are different, and this can have an influence when measuring non-stationary noise.

The voltage spectrum collected with a spectrum analyzer is a time averaged estimate of the spectral components of the incoming signal. This is true for both stationary and non-stationary signals. "Signal" here refers to either recorded signal or media noise - whatever the experimentalist is interested in measuring. For non-stationary noise the only constraint is that the interval of time that the IF is passing through a frequency of interest be long enough to give a valid estimate of time averaged voltage. Practically this means that sweep times should be a long as the experimentalist has patience for. It is important to state that normally the spectrum analyzer does not offer phase or spatial information; and, of course, does not indicate whether the incoming data is stationary or not.

There are other considerations of obtainable accuracy, local oscillator noise, dynamic range, etc. which are important when using a spectrum analyzer. Most of these topics are usually covered in the operating manuals of the given spectrum analyzer. Many of these topics are somewhat more important when measuring well defined spectral features rather than broadly distributed noise. There are some special considerations when measuring noise. Three of these considerations will be discussed here. These include the equivalent noise bandwidth, the effect of passing noise through the IF, and the effect of using a log display. None of these effects change the spectral shape of the noise. However, all of these effects conspire to change the magnitude.

Some examples of noise spectra (collected directly from a spectrum analyzer) are shown in Figure 3. All of these spectra were collected using a generic thin film disk with an MR head. Figure 3 (a) shows the spectrum from the head and electronics. This spectrum was collected with the head lifted away from the disk. For a magnetoresistive or thin film head the spectrum obtained by lifting the head is not discernably different from leaving the head on a stationary disk. However, a ferrite head behaves differently. This is presumably due to eddy current effects. For this reason the ferrite head-only spectrum should be collected with the head loaded on the disk. Spectrum (b) is the total noise from a dc erased track on a film disk. The amplitude is generally less than spectrum (c) which is from a track with a single written frequency. The feature labeled as (d) is the signal peak. The feature labeled as (e) is the shoulder region on the data peak. If a medium has a substantial amount of modulation noise these shoulders usually become much more pronounced. When the integrated noise at a given written frequency is desired the data peak (or peaks) must be suppressed. There is very little discussion in the literature on how this is done. Most investigators just interpolate a straight line under the peak. For a narrow RBW and a medium with low modulation effects the likelihood of significant error appears to be low. There is always a peak at zero frequency which must also be suppressed when integrating spectra.

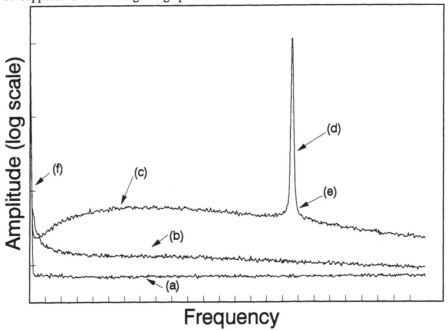

Figure 3. Noise and signal spectra: (a) electronic and head noise, (b) total noise from DC erased track, (c) noise from track with written signal, (d) signal peak, (e) "modulation" shoulders, (f) "DC" peak.

When gaussian distributed white noise is passed through a narrow bandpass filter a Rayleigh distribution is obtained. As an example of the analysis of this class of problems the derivation is given here. Figure 4 shows the analysis diagram.

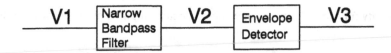

Figure 4. Analysis diagram for Rayleigh distribution.

In Figure 4, V_1 is the input signal plus additive noise; V_2 is the output of the narrow band filter; and, V_3 is the output of the envelope detector. If V_1 is a noiseless sine wave whose frequency is swept through the filter bandwidth, then the spectral shape of V_2 will be the same as the IF filter shape. If V_1 is wideband gaussian distributed noise then V_2 is also dominated by the filter. In this latter case a convenient representation of V_1 is:

$$V_1(t) \; - \; r(t) * \cos(\omega_0 t + \theta(t)) \qquad (4)$$

where r(t) and θ(t) can be considered as independent amplitude and phase random variables. ω_0 is the passband frequency of the narrowband filter. The output of an ideal envelope detector is r(t). In order to calculate the statistics of r(t), it is convenient to first find the statistics of r and θ. A coordinate transformation will be useful:

$$x \; - \; r * \cos(\theta), \quad y \; - \; \sin(\theta) \qquad (5)$$

The joint probability-density functions can now be equated:

$$f_{xy}(x,y)\, dx\, dy \; - \; f_{r\theta}(r,\theta)\, dr\, d\theta \qquad (6)$$

also,

$$dx\, dy \; - \; r\, dr\, d\theta \qquad (7)$$

Since V_1 is assumed to be gaussian and from the assumption of independence of x and y, the joint function can be separated:

$$f_{xy}(x,y) \; - \; f_x(x)\, f_y(y) \; - \; \exp\frac{-(x^2+y^2)/2\sigma^2}{2\pi\sigma^2} \; - \; \exp\frac{-r^2/2\sigma^2}{2\pi\sigma^2} \qquad (8)$$

Therefore, in polar coordinates:

$$f_{r\theta}(r,\theta) = r \exp\frac{-r^2/2\sigma^2}{2\pi\sigma^2} \qquad (9)$$

The density function of just the envelope detector output is obtained by integrating over θ:

$$f_r(r) = \int_0^{2\pi} f_{r\theta}(r,\theta)\,d\theta = \frac{r\exp(-r^2/2\sigma^2)}{\sigma^2} \qquad (10)$$

This is a Rayleigh distribution where σ^2 is the variance of the original gaussian distribution. The mean value of V_3 is then found by:

$$\overline{V_3} = \int_{-\infty}^{\infty} r*f_r(r)\,dr = \frac{1}{\sigma^2}\int_0^{\infty} r^2\exp(-r^2/2\sigma^2)\,dr \qquad (11)$$

The result is:

$$\overline{V_3} = \sigma\sqrt{\frac{\pi}{2}} \qquad (12)$$

The estimated value of V_3 is therefore biased higher (by a constant factor of 1.96 dB) by the behavior of noise as it passes through a narrow band filter. In addition, spectrum analyzers are usually used with a log display. This is accomplished by using a log amplifier. This skews the Raleigh distribution resulting in an additional deterministic loss of -1.45 dB.

One more impact on the absolute accuracy is the fact that noise bandwidths of filters are wider than the bandwidth of the -3 dB points. With some effort this can be calculated but a real filter can deviate from an ideal model. Fortunately it is relatively straight forward to measure the noise bandwidth of the IF by injecting a stable sinusoid, integrating the spectrum and comparing to an ideal brick wall filter. For the spectrum analyzer in the author's lab the measured excess width was equivalent to a +0.63 dB correction (IF = 30 kHz). Manufacturers of spectrum analyzers commonly claim equivalent noise bandwidths corrections of 0.2 to 0.6 dB.

In summary, to use a spectrum analyzer to measure absolute noise amplitudes, a correction must be applied. This correction will have the following budget:

Analyzer scaling	-3.01 dB
Rayleigh distribution	+1.96 dB
Log amp distortion	-1.45 dB
IF noise bandwidth	+0.63 dB
Total correction	-1.87 dB

Thus the measured noise spectrum should be increased by 1.87 dB.

Spectrum analyzers can be set to operate at a single frequency. In this zero span mode each interval is a sample in time rather than frequency. This feature can be used to map the spatial noise distribution on a rigid disk. As an example of using this technique the following medium sample was made to study the angular behavior of signal and noise. A 14" thin film disk with an orientation ratio of 8:1 was fabricated. A 3.5" disk was then milled out the 14" disk. As the head traveled around this smaller disk the orientation ratio varied from 8:1 to 1:8. The signal at 1200 fc/mm (6 MHZ) is shown in Figure 5(a). Two full revolutions of the disk has been plotted. There are two places per revolution where the signal decreases significantly. These locations are where the easy axis is orthogonal to the head direction. The corresponding noise information is shown in Figure 5(b). This is a zero span trace taken at 1.8 MHz (in the presence of the signal). The signal information (Fig. 5a) could have been adequately measured with an oscilloscope. However, the spectrum analyzer in zero span mode is the easiest method of measuring the spatial distribution of noise.

4. Time Domain Measurements

4.1 Sampling Oscilloscopes

A sampling or digitizing oscilloscope is one which converts an incoming analog waveform and stores the result as digital information. For some applications the sampling scope is a very useful instrument. Fortunately, in the past few years the capability of these instruments have improved substantially offering researchers new measurement opportunities. There are three aspects of a sampling scope which determine the performance of the scope. First, the speed with which repetitive samples may be taken is important. Commercially available oscilloscopes are presently able to sample up to 2 GHz (2×10^9 samples per second). The second consideration is the number of intervals available from the analog-to-digital converter (ADC). Eight bit ADCs are common - these give a range of 256 voltage intervals. And finally the depth of memory is important. The memory depth is the number of samples which can be stored in memory. This, of course, depends on the amount of fast memory in the instrument. As the cost of high speed memory devices continue

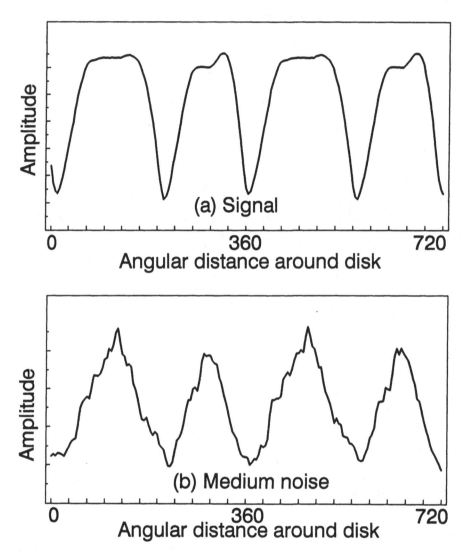

Figure 5. Zero span measurements of signal (a), and medium noise (b).
See text for details of medium.

to fall the memory depth of sampling scopes can be expected to increase. Memory depths of 64k are now readily available. There are many other features and attributes contemporary sampling scopes have which influence the performance and convenience of use. Manufacturer's published literature on these scopes is an excellent source of information.

Most studies of noise with sampling scopes involve making many measurements of the same set of readback pulses in order to "average out" the electronic and head noise (the stationary component). In any conversion from an analog signal to a digital record some information is lost due to quantization noise. This is an important consideration so an expression will be developed here to quantify the quantization noise.

Assume an incoming signal spans a voltage range of V_m. V_m could be considered the peak-to-peak amplitude of a readback signal. Assume also that Δv is the voltage per quantized level. Therefore, there are N levels over the range of V_m:

$$N = \frac{V_m}{\Delta v} \qquad (13)$$

Analog voltages in the range of $V_i \pm \Delta v/2$ will be given the digital value of V_i. Thus the error term will have a maximum value of $\pm \Delta v/2$ assuming the ADC made the best decision. If the analog voltage can be represented by the digitized voltage plus an error term, then:

$$V_{actual} = V_i + e, \qquad -\Delta v/2 \le e \le \Delta v/2 \qquad (14)$$

then the rms error can be calculated by:

$$E(e^2) = \frac{1}{\Delta v} \int_{-\Delta v/2}^{\Delta v/2} e^2 de = \frac{\Delta v^2}{12} \qquad (15)$$

Therefore, the rms error, n_q, is:

$$n_q = \frac{\Delta v}{2\sqrt{3}} \qquad (16)$$

where n_q can also be called the quantization noise. It is useful to consider a signal to quantization noise ratio:

$$\frac{S}{n_q} - \frac{V_{o\text{-}p}}{\Delta v/2\sqrt{3}} - \sqrt{3}\, N \qquad (17)$$

where $V_{o\text{-}p}$ is the base to peak amplitude. Thus the only way to improve the quantization S/N is to increase the number of utilized levels. This can be done by using an ADC with as much resolution as possible. Also, the gain can be manipulated to use as much of the full range of the ADC as possible.

As a numerical example if a signal filled half the range of an eight bit ADC the quantization ratio would be:

$$S/n_q - \sqrt{3}\, 128 - 47 \ dB \qquad (18)$$

The medium or system S/N would have to be lower than 27 dB in order for the quantization noise to be less than 10% of the total noise.

The discussion above applies to a single record capture of data. In many cases it is desirable to capture many traces and then average to reduce the non-repeating component. In these cases the synchronization of the trigger signal within a timing interval can effectively add more quantization noise. One case to consider is where the trigger source is much noisier than a timing interval. See section 4.2 for a discussion of the relationship between noise and jitter. Noisy triggering can be reduced to some extent by a de-jitter procedure. This involves shifting successive records to obtain the minimum least squared error. Generally the jitter can be reduced to one time interval. Then, if it is equally likely to be anywhere within a timing interval then the rms timing error can be calculated as follows:

$$t - t_i + \tau, \qquad -\Delta t/2 \leq \tau \leq \Delta t/2 \qquad (19)$$

$$E(\tau^2) \ - \ \frac{1}{\Delta t} \int_{-\Delta t/2}^{\Delta t/2} \tau^2 \, d\tau \qquad (20)$$

Therefore,

$$t_q(rms) \ - \ \frac{\Delta t}{2\sqrt{3}} \qquad (21)$$

This value of timing noise can be converted to equivalent voltage noise by multiplying by the local signal slope:

$$n_q = t_q * V_i \qquad\qquad V_i = \frac{V_i - V_{i+1}}{\Delta t} \qquad\qquad (22)$$

There can be occasions where this noise term is appreciable. Faster sampling times will help reduce it.

It has been assumed here that there is negligible clock jitter within the sampling scope. For contemporary scopes this jitter is very small, a few picoseconds, and is difficult to measure.

4.2 Time Interval Analysis

Most digital recording systems depend on extracting a sequence of events in time in order to read the stored information. Usually the events are the times of crossovers of differentiated read back pulses. If the time of crossover is shifted due to noise induced jitter or interference, then the error rate will degrade. Thus the measurement and analysis of timing intervals between subsequent time related events is a very useful and relatively direct method of evaluating system performance.

In the past few years several vendors have offered instruments capable of measuring distributions of time intervals (Strassberg, 1991). Some of these instruments need a small interval of time to store the information from a single measured interval. As a consequence not every interval in the recorded stream is captured. However, the capability (speed and memory depth) of these instruments seem to be improving very rapidly with each new model, so this sampling problem will become less of a problem.

As mentioned above the structure in the distribution of time intervals is due to jitter and various kinds of interferences. The contribution of noise to measured jitter distributions is relatively straight forward to analyze if the noise is gaussian distributed (Nunnelley, et al., 1987). If the signal is a sine wave and the noise is gaussian distributed then the crossover jitter is also gaussian and has a width of:

$$t_{rms} = \frac{1}{2 \pi f_s \, S/N} \qquad\qquad (23)$$

For a more general signal the result is:

$$t_{rms} = \frac{N_{rms}}{signal\ slope} \tag{24}$$

Interferences can happen for a variety of reasons including details of the magnetic interface (pulse crowding, etc), pattern sensitivity, and adjacent track read sensitivity, *inter alia*. These effects require more effort to account for analytically and direct time interval analysis may be one of the best ways to study these effects.

5. Certification Techniques

The purpose of a manufacturing screening procedure is to insure that a manufacturing process is stable. Ideally the screening procedure is used in real-time process control so that a given parameter is forced to be narrowly distributed and has a time invariant mean. Generally these certification techniques should be very rapid and repeatable.

Noise from particulate media is stationary and largely independent of the presence of recorded signal (for a medium with good dispersion). As a consequence most certification noise measurement schemes involve measuring the noise (after a dc erasure) with a broadband power meter. This technique is fast, accurate, and does not require much data manipulation.

Noise from thin film media is not stationary and the averaged noise is highly dependent on recording signal density. There are two techniques in common use. The first involves writing a relatively high frequency (usually the highest data frequency) and using either a notch or low pass filter to remove the signal component. The resultant noise is then presented to a broadband power meter. The noise values obtained with this technique are somewhat optimistic since some noise is filtered along with signal. However, in general, the techniques give reliable relative numbers. The second method involves reverse dc current erasure (Pressesky and Lee, 1990). First the track is dc erased. Then the dc current in the opposite direction is incrementally increased and the broadband noise at each increment is measured. The current which corresponds to the maximum noise can be calibrated to Hc. The peak noise values obtained by this technique should be used with some caution. If there are Hc variations around the track then the noise at a given current will be for only part of the track. This problem could be solved by integrating noise with respect to reverse current. The other problem is that domain structure for a reversed dc erased track may be different than for a track with written transitions. This appears to be especially true for highly oriented films.

6. Miscellaneous Techniques

6.1 Track Edge Noise

Generally any magnetic boundary recorded in a thin film medium results in a boundary with zigzags patterns and therefore noise. Noise in written data transitions along the track (perpendicular to the head direction) is of primary interest. However the track edge is another boundary which has some random magnetic structure and noise. Track edge noise is particularly important in system considerations where disturbances in the servo system might be expected. It is possible to measure the track edge noise by a technique developed by E. Yarmchuk (Yarmchuk, 1986). The technique is based on a time domain sampling procedure. A triggered segment of a track is recorded for thousands of passes and these records are averaged to reduce the electronic and head noise. Then the head is moved radially a small fraction of a trackwidth and another highly averaged waveform is collected. The difference between these two waveforms is equivalent to two heads of much more narrow trackwidth and opposite polarities. The effective narrow trackwidth can then be used as a spatial probe with very high resolving power. With many radial increments an image of track edge noise can then be compiled. This procedure is labor and computation intensive but offers detailed information on the structure of track edge noise. Track edge noise measurement can be measured with a spectrum analyzer (also discussed by Yarmchuk, *ibidem*). This appears to be a more convenient measurement but offers less detail on the spatial distribution of noise.

6.2 Overwrite Measurements

Overwrite "noise" is not an intrinsic noise source but depends on such system parameters as magnetic spacing, head geometry, write current, etc. Perhaps overwrite should be considered physically as an interference but it can be treated analytically as an additional noise source (E. Williams, this volume). The customary method of measuring overwrite is similar to the measurement of media noise discussed above. First a low density pattern, usually equally spaced transitions, is written. This is then overwritten with a higher density pattern. The residue of the lower frequency is then recorded. This measurement is usually made with a spectrum analyzer although filtering techniques can also be used. This basic technique has been extended to include measurements of the spatial distribution of overwrite across a track (Palmer and Peske, 1990).

7. Acknowledgements

Technical assistance and helpful discussions with Mark A. Burleson and Dan S. Parker during the preparation of this chapter are gratefully acknowledged. The support and assistance of Christine H. Smith is greatly appreciated.

8. References

The manufacturers of the instruments described in this chapter publish manuals, sales brochures, and applications notes which are generally very useful. The cited reference from Hewlett-Packard is especially informative about spectrum analyzers. There are several books available which discuss analytical topics related to noise measurements. The cited work by Schwartz covers most of the issues in an accessible style. The work by Papoulis covers some of the same material at a more advanced level.

(Anonymous), "Spectrum Analysis Basics", Application Note 150, Hewlett-Packard Co., 1989.

R. A. Baugh, E. S. Murdock, and B. R. Natarajan, "Measurement of Noise in Magnetic Media", *IEEE Trans. Magn.* **MAG-19**, p. 1722, 1983.

Julius S. Bendat and Allan G. Piersol, *Engineering Applications of Correlation and Spectral Analysis*, (John Wiley & Sons, 1980).

Morris Engelson, "Development of the Modern Spectrum Analyzer", *MSM*, p 47, October 1989.

John Gallant, "Spectrum Analyzers", EDN, p 95, May 9, 1991.

J. B. Johnson, "Thermal Agitation of Electrons in Conductors", *Phys. Rev.* **32**, p. 97, 1928.

L. Nunnelley, R. D. Harper, and M. Burleson, "Time Domain Noise-Induced Jitter: Theory and Precise Measurement", *IEEE Trans. Magn.* **MAG-23**, p. 2383, 1987.

Dean Palmer and James V. Peske, "Spatial Distribution of Overwrite Interference on Film Disks", *IEEE Trans. Magn.* **MAG-26**, p. 2451, 1990.

Athanasios Papoulis, *Probability, Random Variable, and Stochastic Processes*, 2 nd Ed., (McGraw-Hill, 1984).

J. L. Pressesky and S. Y. Lee, "In Situ Measurements of the Coercive Force on Thin Film Discs - A comparison of Different Techniques", *IEEE Trans. Magn.* **MAG-26**, p. 2472, 1990.

Mischa Schwartz, *Information, Transmission, Modulation, and Noise*, 3 rd Ed., (McGraw-Hill, 1980).

Dan Strassberg, "Frequency and Time-Interval Analyzers", *EDN*, p.79, June 6, 1991.

Edward J. Yarmchuk, "Spatial Structure of Media Noise in Film Disks", *IEEE Trans. Magn.* **MAG-22**, p. 877, 1986.

INDEX

276